COMPROMISED

COMPROMISED

Counterintelligence and the Threat of Donald J. Trump

PETER STRZOK

HOUGHTON MIFFLIN HARCOURT

BOSTON   NEW YORK

2020

For information about permission to reproduce selections from this book,
write to trade.permissions@hmhco.com or to Permissions, Houghton Mifflin Harcourt
Publishing Company, 3 Park Avenue, 19th Floor, New York, NY 10016.

hmhbooks.com

ISBN 978-0-358-23706-8

Book design by Chloe Foster

Printed in the United States of America
DOC 10 9 8 7 6 5 4 3 2 1

For my family

Contents

PART III

A free people ought to be constantly awake, since history and experience prove that foreign influence is one of the most baneful foes of Republican Government.

— GEORGE WASHINGTON, FAREWELL ADDRESS, 1796

In our country, the lie has become not just a moral category, but a pillar of the State.

— ALEKSANDR SOLZHENITSYN, 1974

Author's Note

In keeping with the FBI's practice in public documents, I do not use the true names of FBI employees who were not or are not in senior leadership positions. Their characters are real, but their names are not.

The views expressed in this book are my own and represent my opinions based on my current understanding of the facts, my experience, and my observations.

Introduction

EARLY IN 2017, shortly after the inauguration of the 45th president of the United States, the top leadership of the FBI's Counterintelligence Division filed into a small room on the fourth floor of the J. Edgar Hoover Building. A weak midwinter light shone through the wide plate-glass windows overlooking Pennsylvania Avenue. The room, tucked away off a corner of the Bureau's counterintelligence front office, was reserved for high-level briefings and especially sensitive debates. Meetings like the one that was about to start.

I took a seat opposite the door, facing a long whiteboard, about eight feet wide and four feet high, which occupied one side of the room. Across the broad surface, names were carefully written in erasable marker. At the top, I read the initials "DJT" written in small blue letters. Below that were names familiar to all of us in the room, some of which soon would become known to the American public. Paul Manafort. George Papadopoulos. Mike Flynn. Carter Page.

All the names on the whiteboard belonged to people inside the Trump administration or connected to it. All had something in common: we had received credible counterintelligence allegations against each individual. We had already opened investigations into some of them. There were other names on the board as well, belonging to people for whom we didn't have open cases, although by regulation we had more than enough evidence to do so. Opening those cases was the subject of ongoing debate — a back-and-forth that sometimes grew heated.

It was a very difficult time, and we still had many more questions than answers. On that day we grappled with an especially troubling question, one that none of us could have anticipated in our wildest imaginations: whether to open a counterintelligence case against the president himself.

Over the course of the previous year's presidential campaign, and in the aftermath of the 2016 election, concerns within the FBI's Counterintelligence Division about Trump and his advisers had grown steadily into outright alarm. In late December 2016, the director of national intelligence published a report confirming intelligence we had been receiving throughout the presidential campaign: that Vladimir Putin and the Russian government were interfering with our electoral system to undermine faith in the election, hurt Hillary Clinton's prospects, and help Trump get elected.

Those of us gathered in the room at FBI headquarters that day also knew that the Russians had pulled some of their punches; we had credible intelligence that Russia possessed the means to have done even more damage to our electoral system in 2016 but had held back. The knowledge that they had something in reserve to potentially use against us in the run-up to our next presidential election only made the atmosphere in the small conference room that much tenser.

The stress and uncertainty of the past several months weighed heavily on all of us. I was proud of how our team had responded to the pressure. We maintained our professionalism and confidentiality, keeping a quiet discipline as we went about the unprecedented task at hand. Still, a grim stoicism hung over our work. There was little levity; this was nothing to joke about. We never discussed how it affected us personally, but we shared an unspoken sense of taut apprehension over the enormity of the events before us. There was a silent acknowledgment: *I know, I feel it too. Let's keep driving forward.*

For many months before this moment, I had also been immersed in another politically delicate case. Since August 2015, I had supervised the FBI investigation into whether Hillary Clinton had mishandled classified email during her tenure as secretary of state. Toward the end of that investigation — which culminated in a weekend interview with Secretary Clinton just before July 4, 2016 — I sat with the core group of our team to review our findings. One by one, I asked each member if he or she thought there were any remaining investigative threads that might change our fundamental understanding of the case. And then I asked them, one by one, whether they were comfortable with the decision not to recommend criminal charges. The answer was unanimous and unequivocal: we had done a complete and exhaustive investigation, and the facts did not support bringing charges over the email.

But Trump was a totally different matter.

By the time Trump was sworn in, the FBI had been investigating Russian election interference for almost six months — and what the Bureau knew about the cases was far different from what the public did. The public knew that the FBI had investigated Clinton, but even within the Bureau very few people were aware of our investigations into Russian interference. And if the American people had known what we did at the time of the election, they would have been appalled.

For starters, we knew that one of Trump's foreign policy advisers, Papadopoulos, had boasted about his knowledge that Russia had information damaging to Hillary Clinton and Moscow had offered to assist the Trump campaign by releasing the material. A week after the inauguration, the Bureau interviewed Papadopoulos, and he lied to us. Despite telling us during a second interview a few weeks later that he'd cooperate, he left the interview and deactivated a Facebook account that he had used to communicate directly with Russians.

That was just the tip of the iceberg. We also knew that Attorney General Jeff Sessions had attended campaign-related meetings with Russian officials, even though he later denied having had any such contact when questioned during his Senate confirmation hearings. We knew that National Security Advisor Mike Flynn had had conversations with Russian officials, then covered it up; he later swore to two judges on two separate occasions that he had lied to me and another special agent in a White House interview about those conversations. We knew Trump campaign chairman Paul Manafort was in deep financial trouble and had a tangled web of connections to pro-Russia Ukrainians and Russian intelligence-related personnel, something unheard of in a presidential campaign manager. We knew that Trump's history of financial relationships with Russia going back to the 1980s was completely at odds with statements he made to the American public on TV and on the campaign trail.

Trump's pronouncements about Russia had become almost a bad counterintelligence joke. Instead of accepting the U.S. intelligence community judgments that Russia was engaged in active measures to undercut our election, he sowed doubt with a false narrative that perhaps an obese mystery man had hacked Democratic Party networks. Instead of acknowledging that he had business interests in Russia and that his fixers had been in the country's capital trying to negotiate a deal for a Trump Tower Moscow, he denied having any ties to Russia. Every call to Russia to hack Clinton's email, every speech praising WikiLeaks for releasing material stolen by the Russians, each question about the U.S. commitment to the NATO alliance, every false moral equivalence between the U.S. and Russia — they all lined up with Russian strategic interests and followed a script that the Kremlin would have been hard-pressed to write more effectively.

A skeptic might point out that as far as that behavior relates to Trump, none of it is necessarily criminal. That's true. But that's also not the point. For any intelligence agency, one of the top strategic goals is to understand an adversary's actions and develop ways to influence that behavior, to give that agency's government a competitive advantage. Russia does that better than most other countries, and it does it in a myriad of ingenious ways, including exploiting ideological, financial, and other coercive vulnerabilities of people able to help it achieve its ends. The aim of counterintelligence is not necessarily to build a criminal case against these manipulated individuals, or prove that a law has been broken; it's to understand how they are contributing to an adversary's intelligence needs and to apply that knowledge toward the ultimate goal of protecting the United States and advancing our own strategic interests.

Trump, who had turned to the Russians for financial assistance in his failing business enterprises as early as 1987, had just such exposure. And from the perspective of the FBI's Counterintelligence Division, many of Trump's actions, large and small, public and secret, were so inexplicably aligned with those of Russia that the coincidence — if it was a coincidence — had become impossible to ignore.

This urgent issue is what had brought together the leaders of the Bureau's Counterintelligence Division in this secluded fourth-floor meeting room. We needed to ask a question that had never before arisen in the entire 240-year history of our republic: whether the president of the United States himself might be acting as an agent of a foreign adversary.

Given the fountains of vitriol poured upon me since 2018, it may come as a surprise that at the time I was not in favor of opening an investigation on Trump himself. There were both practical and philosophical arguments against it. From a tactical perspective, how do you conduct surveillance of the president? You can't quietly ask the Secret Service to collect his trash or tail his motorcade or review his financial records without attracting attention. From a strategic perspective, such an investigation would require notifying the Department of Justice (DOJ) and Congress, and those discussions would never remain secret in the Washington fishbowl, where political appointees occupy the uppermost echelons of the departments of the federal executive branch, and where partisanship seemed to have replaced patriotism as the guiding light for many of these ostensible public servants.

Moreover, at that stage, and in light of all the investigations we already had open, I didn't think we needed to open a case against the president himself to effectively assess the Russian interference in our elections. The existing cases, I felt, were sufficient to enable us to defend national security. I argued that we could always revisit the decision if further information came to light.

From a legal and constitutional perspective, furthermore, the Bureau had never faced a situation like this. How would we go about investigating a sitting president as a possible national security threat, if that became necessary? The president sets American foreign policy, and as the highest elected official charged with protecting the nation, can he be at odds with U.S. national security? And should it be up to the FBI, operating in the long shadow of former director J. Edgar Hoover, to investigate that? But if not the FBI, then who, especially in a way that would maintain the ability to credibly investigate if such a decision were made?

As we debated these and other questions, we all knew the constitutional parameters that both empowered and restrained us as employees of the federal government's main law enforcement agency. We all knew that the Constitution grants Congress the ability to impeach and convict "the president, the vice president, and all civil officers of the United States" if it believes they are guilty of "treason, bribery, or other high crimes and misdemeanors." But that power only removes the person from office; Congress neither conducts criminal investigations nor prosecutes those crimes. That's what the FBI does, if necessary.

Yet merely imagining a counterintelligence case on the president, never mind the prosecution that could potentially result from such an investigation, took us far outside the bounds of established procedure, and far beyond the scope of most legal textbooks. Duty required those of us huddled on the fourth floor of the J. Edgar Hoover Building on that winter day to struggle through a thicket of questions that no one at the Bureau had ever needed to ask, let alone answer. As we formally entertained the notion that the president could be a witting or unwitting pawn of Russia, we were forced to wonder how we could do our jobs while assuring that we worked within the letter and the spirit of the law. We pondered how to protect the country while best maintaining the public's faith in the FBI and our system of government if this terrible eventuality ever became reality.

Our oath of office, our duty to protect and defend the Constitution, was not an abstract idea that afternoon. It was an urgent imperative — one born, in my case, out of a keen sense of the fragility of democracy, a lesson learned at a very young age.

A CITY UPON A HILL

Smudges of ash and burning paper floated up into the cold air, glowing flakes of obliterated words carried into the twilight in the hot updraft of a roaring bonfire. My father stood next to a 55-gallon drum in a rubble-filled lot adjacent to our walled home in the foothills outside Tehran. He held a stack of file folders in one hand and fed them into the flames with the other.

It was December 1978, and Iran was coming apart. In the waning light of dusk, my dad was burning documents that were better destroyed than left behind. The fire in the metal drum licked upward above the rim, leaping as he added fuel. I watched as the heat from the open flame sent the smoldering documents whirling high up into the sky, lifting them out and away from the flame in a geyser of sparks and half-burned paper. My job was to chase down wayward bits of unburned paper that floated back to earth and return them to the fire, a task probably intended to consume my eight-year-old's energy as much as fulfill an actual need to ensure their destruction. Over and over I dashed across the lot in pursuit of the charred pieces of my father's records, brought them back, and carefully returned them to the fire.

I have a dim memory that we were not alone. Off to one side, on the street at the edge of the lot, a man stood watching us, more enigmatic than menacing. He was far enough away that it was impossible to see details about him, and the distance of memory has made that recollection even more indistinct. I don't know if his presence was by chance or by design, but his attention was intense. He was silently observing the destruction of those life and work records judged imprudent to keep amid the gathering storm of revolution.

Years later, as an FBI agent, I would learn the power of stillness. During an interview, silence can lead people to feel great disquiet. Uncertain about what their interlocutor wants or is thinking, people sometimes try to fill that gap, creating purpose and explanation and meaning without any factual basis, or laying bare the truth that they had hoped to keep hidden. But in that vacant lot, with the lights of Tehran coming on and the distant Mount Damavand fading into the darkening sky, my young self had yet to develop a deep understanding of moral conflict or ambiguity. The man stared at me, and I stared back at him in the approaching Persian dusk.

At the end of 1978, the prospect of revolution was no longer editorial-page conjecture. Discontent and violence rocked the country day after day. A strike among oil workers in early November had shut down the oil industry. Rioting followed, with furious mobs burning and ransacking government offices, foreign embassies, and other buildings. The shah responded by imposing martial law. In December, violent clashes broke out nightly between the army and demonstrators, with the soldiers firing live bullets into the crowds. The BBC estimated that 700 people were killed. Foreigners began to flee.

After weeks of valiantly trying to protect its school buses with adult escorts and chicken wire over windows to deflect thrown rocks, the Tehran American School had shuttered its doors. My winter break was suddenly indefinite. I spent it watching order unravel. When we drove, my father, recently retired from the army and now working for Bell Helicopter, would cut the car engine and coast downhill to save fuel, which was suddenly rationed through hours-long gas station lines, a victim of supply disruption in oil-rich Iran. At night I listened to the sound of gunfire and tanks growling through the streets. When the sun rose in the morning, we would pass still-smoldering debris from the previous night's violent protests, pushed to the side at intersections.

Banks, theaters, and other symbols of "Western oppression" were attacked, burned, and abandoned. The sweetly acrid smell of kerosene lamps and heaters settled about our house, a byproduct of nightly shutdowns of power as part of the government-imposed curfew intended to maintain control. My mother and I were able to board a flight out of Mehrabad Airport in early January, just days before the shah fled the country and a new kind of darkness settled over Tehran. I remember pressing my face to the window as the airplane banked away from the Iranian capital and gazing down on the sparkling yellow lights of the city sprawled along the foothills of the Alborz Mountains before the aircraft turned and I lost sight of Tehran forever.

American views of Iran today remain deeply rooted in the anti-American image of Ayatollah Khomeini's Islamic Revolution and the seizure of American hostages in 1979. The invasion of the U.S. embassy inflamed anti-Iranian fervor in the United States to a fever pitch, and for some, that fury still smolders today. But my experience in Tehran taught me to appreciate complexity. I saw televised images of crowds chanting "Death to America!" and militants leading blindfolded American hostages through the U.S. embassy compound. I also saw the faces of friendly Persian friends and neighbors: kind, highly educated, cultured, and unmatched in their hospitality. Even as I remember the stench of violence and the blanket of fear across Tehran, I also recall the acts of generosity and fellowship from those we left behind as our plane took off, refueling in Kuwait before continuing on to London. Having survived and escaped the Iranian revolution, I still smile inwardly at the memory of the many good people I met.

But complexity doesn't mean there are no absolutes. As a child, I knew there was such a thing as good and bad in the world. This view has only grown stronger in me over the years. The cliché is that wisdom comes with age, and experience brings a greater appreciation for nuance, a tolerance for shades of gray. For me, age and experience have brought the opposite. Now more than ever I see the world in clear delineations of right and wrong, the latter as different from the former as night is from day.

This might seem like a contradiction. As a law enforcement agent with over two decades of experience in counterintelligence, I have had the job of carefully examining each problem from every perspective, of analyzing intelligence from all angles, from inside and out, taking into account all points of view. But at the end of the day I must make a judgment about the facts or fictions before me, however murky or cryptic or illusory they may be. I have come to believe that truth and lies, black and white, innocence and criminality, can be clearly defined. Truth can be devilishly opaque, and can't always be explained in five minutes, or even in 500 minutes, let alone 280 characters. But I know this: truth is ultimately knowable and incontrovertible.

Leaving Iran didn't end my education about the strength and stability of American democracy. Having retired from the army, my father had a brief stint in the contracting world before starting a second career in international development work. After a year in Virginia, while he worked in Washington, we moved in 1980 to Burkina Faso (then called Upper Volta) in West Africa. If Iran's upheaval was a slow-simmering revolution decades in the making, Burkina was in a constant hot boil of violent unrest. Over the three years we were there, the government fell in two coups d'etat, with several more unsuccessful attempts. The immediate aftermaths of the coups were depressingly predictable: the plotters installed a puppet loyalist to head the military and national police, sealed the borders, seized the airport, shut down newspapers and television and radio stations (instituting, as often as not, a 24-hour loop of martial music), and imposed a dusk-to-dawn curfew. This was so invariably formulaic because authoritarianism demanded it: seizing the key levers over the government and populace was the only way for would-be dictators to gain, consolidate, and maintain power.

Reliable local information was scarce in Burkina Faso, so we turned to the shortwave radio in the living room for news. While it might sound exotic, it wasn't. In the days before the Internet, information was easier to control and manipulate; most expatriate families relied on international broadcasts, Voice of America or the BBC for news and the Armed Forces Radio and Television Network for stateside sports. Our radio, the size of a small toaster oven, required a wire antenna tacked on the wall. Consulting a homemade chart with times and frequencies, I'd turn the radio's big black knob to tune the glowing green fluorescent display to whatever frequency carried the news at that hour. It was an ingrained nightly ritual, but one we adhered to especially during periods of political unrest. To this day I cannot hear the "Lilliburlero," the BBC theme music at the time, without an inner voice appending five short tones and one long tone, followed by an English-accented "18 hours, Greenwich Mean Time. BBC World Service, the news, read by" a Very British Name.

A few years later I repeated the process, only this time listening to broadcasts much closer to home as Haiti fought to emerge from decades of dictatorial rule by the Duvalier family, whose oppressive regimes ended with a proud nation struggling to overcome crippling poverty while building a nascent democracy. By that time I was old enough to understand the more brutal aspects of dictatorial rule. Just as the shah of Iran had wielded SAVAK, his domestic security agency, as a weapon against his own people, so too did François Duvalier (and later his son, Jean-Claude), with a brutal secret police force known as the Tonton Macoute. Named after a bogeyman from Haitian lore, it was used by the Duvaliers to suppress political opposition and punish — with imprisonment, torture, and execution — those who dared challenge them, even if only with words.

By the time I entered college, I had lived through four revolutions on three continents. Whether in Iran, West Africa, or Haiti, all shared common characteristics, and all taught me lessons about dictators and authoritarians and their hunger to consolidate power and obtain — or at least convey — legitimacy. That quest for legitimacy played out in a host of ways. One was the desire to manipulate, control, or discredit media. A relentless distortion of reality numbs a country's populace to outrage and weakens its ability to discern truth from fiction.

Another way dictators sought to secure power and legitimacy was by co-opting the power of the state — its military, law enforcement, and judicial systems — to carry out personal goals and vendettas rather than the nation's needs.

Still another was by undermining dissent, questioning the validity of opposition and refusing to honor public will, up to and including threatening or preventing the peaceful transfer of power.

These were lessons that I would learn again and again in the years to come, but always in distant lands, always in places that lacked the robust democratic institutions of the United States and the legal scaffolding that supports it, the Constitution. I never thought I would have occasion to revisit these lessons at home, in the United States. And I never expected see the grotesque traits of dictators in Haiti or Iran reflected in my own commander in chief.

TRUE FAITH AND ALLEGIANCE

When I was 17, I followed my father's footsteps into the army by entering the Reserve Officers' Training Corps. After graduating from the field artillery officer basic course in Fort Sill, Oklahoma, I was assigned to the 101st Airborne Division at Fort Campbell, Kentucky. There I was embedded in an infantry company in the 502nd Infantry Brigade, which traced its proud lineage to World War II, its D-Day heroics memorialized in a famous picture of General Eisenhower speaking to its paratroopers before they loaded onto airplanes to cross the English Channel to liberate Nazi-occupied France. As I was promoted into various leadership positions within one of the division's artillery battalions, we maintained a demanding cycle of training, a deployment period when we were subject to 18-hour recall to be airlifted anywhere in the world, and an administrative recovery period, all of which repeated throughout the year.

The army continued my democratic education, reinforcing in me the bedrock principle of fidelity to the Constitution, adherence to the ethical laws of war, and respect and support for civilian control of the military. I had earnest debates during officer professional development sessions, a type of educational training seminars within units, about whether or not U.S. Army officers should even vote in presidential elections. The argument against doing so was that we were bound by the decisions of the duly elected president, and having a preference or stake in that outcome by voting — let alone campaigning in uniform, which was forbidden — might unduly prejudice our ability to serve that president, who held the authority to commit us and the soldiers we led to war.

While I served on active duty, I didn't agree with the view that officers should abstain from voting, and I don't agree with it now. Having seen so many disenfranchised people around the world, I understood voting to be at the core of our democracy. Honest service to the nation need not require being a political eunuch; one should be able to vote, then have the strength of character to support the outcome, whatever it is. These principles, the twin pillars of democracy and duty, are what gives America its strength, and they have guided me and millions of others throughout our military and public service.

Shortly after I entered the military, I turned 18 and gained the right to vote. It may surprise some to learn that I grew up in a Republican household, and like many in the FBI and our armed services, I have frequently voted for politicians of that party. But while I place great value on our law enforcement, military, and national security strength, I have always cast my ballot for the candidate who I believed would best protect our nation, priding myself on voting based on all the evidence available to me rather than simply out of partisan loyalty.

The army paid my way through Georgetown University with an ROTC scholarship, and as my service commitment came to an end in 1995, I knew that I wanted to continue in government service. In April that year, an ex-army soldier named Timothy McVeigh drove a rented van loaded with an improvised explosive device up to the federal building in Oklahoma City, parked, and walked to safety before the bomb went off. Its devastating detonation killed 168 people and focused the nation on the threat of homegrown domestic terrorism. Congress responded in part by appropriating funds for 60 new FBI counterterrorism analysts. Seeing a job advertisement in the paper, I applied. It was the best career decision I ever made.

I quickly fell in love with the complexity and tangible impact of the work we did at the FBI, the near-universal passion of my colleagues, and our shared belief in our critical role in protecting the American people. Within months I knew I wanted to be an agent.

So began my two-decade career in counterintelligence, fighting a host of foreign threats but especially ones from the Russians and Chinese. Their goals and methods were very different. The Russians, like the Americans, viewed national security from a very Western perspective of balancing powers and military strength, whereas China was deeply engaged in targeting an even more fundamental source of power: our economic might. Yet for much of this time, the bulk of our work against China was the same as it was against Russia: tracking their intelligence agencies' attempts to enlist the help of our fellow Americans, either wittingly or unwittingly, in their shadow war against our nation. During this time I participated in many cases that remain classified, as well as one high-profile case that would form the basis for a popular TV show. Yet in most of these cases the tactical fundamentals were not significantly different than they had been during the Cold War. Not until the Internet revolution at the turn of the century did the Bureau enter the phase of the counterintelligence struggle that we find ourselves in today, in which spies conducting dead drops and using invisible ink commingle with social media bots waging elaborate online disinformation campaigns aimed at destabilizing our democracy in new and frightening ways.

The FBI was the job of a lifetime. I loved working there, in large part because of the shared patriotic purpose I felt when walking in the door each morning. Going to work each day also felt like sitting down in front of an enormously complex and satisfying puzzle. I got to make a career of pursuing the truth, alongside some of the best people I've ever met.

About a year after the events described in this book, I spoke at a class at Harvard Law School and answered a number of really challenging questions. One of the students asked if, looking back, I would do anything differently in my investigative work. Of all the questions, that might have been the easiest to answer.

No, I answered almost immediately. We worked hard and had made America a safer place. It was an honor to have served with the men and women of the FBI, and although I wish things had turned out differently, I wouldn't change a minute of any of the work we did.

CORRUPT INCOMPETENT HATING FRAUD

I have devoted my adult life to defending the United States, our Constitution, our government, and all our citizens. I never would have imagined — I could not have imagined — that the president of the United States, the most powerful man in the world, would single me out with repeated attacks of treason, accusing me of plotting a coup against our government. Suggesting that I be executed. Labeling me "corrupt," "incompetent," a "sick Loser" and a "hating fraud." And, of all things, questioning my patriotism while he stood in Helsinki next to Russian president Vladimir Putin, an adversary who has been actively attempting to subvert democracy in the U.S. and around the world.

I also never could have anticipated writing a book about counterintelligence, let alone one about the counterintelligence threat posed by the Trump administration. I didn't *want* to; most intelligence professionals wouldn't.

There are reasons for that reticence. One is simple: our work is sensitive and usually classified. We operate quietly; we shun attention that would jeopardize our investigations. With such limitations, it is difficult and frequently unwise to tell a complete and persuasive story.

I didn't seek the spotlight; it found me. After the gist of text messages that I sent expressing my personal political views was cynically leaked to the *New York Times* and *Washington Post* (likely by a high-ranking official at the Department of Justice or White House; more on that later), my days of being anonymous were over; my cover as a counterintelligence agent was, in a very real sense, blown.

Today, moreover, I don't feel that I have the option of keeping quiet about the clear and present danger that I know the Trump administration poses to our national security. Whether I like it or not, circumstance has put me in a position to address the challenges that the nation faces in these conflicted times, and I feel it is my duty as an intelligence professional and an American to do so.

My upbringing and my love of my country propelled my career in counterintelligence. Russia's attack on our elections and Trump's and his campaign's willingness to accept Russian help — including a pattern of behavior that ultimately led me and my colleagues at the FBI to launch an investigation of the president — were shocking and repugnant to me as an American. The things I saw during that time prompted such alarm for me and my colleagues that I now feel it is my duty to share with the public what I can of those events.

Let me be clear. I have made mistakes, and I have paid a high price. I deeply regret casually commenting about the things I observed in the headlines and behind the scenes, and I regret how effectively my words were weaponized to harm the Bureau and buttress absurd conspiracy theories about our vital work. I never foresaw the unprecedented way the nation's political dialogue would be twisted — how traditional boundaries would be cast aside, character assassination would become commonplace, and an all-out effort would be made by partisans up to and including the president of the United States to destroy the credibility of the institution I love so dearly.

Most importantly, I should never have allowed myself to make the worst mistake of my life, which harmed my family and my wife, the love of my life since I met her almost 30 years ago. I am ashamed of having done so and determined to make things right with them. This is why you will not be hearing much in this book about my personal life. My family has suffered enough because of my actions; they deserve their privacy, and the space to heal from the countless traumas they have endured over the last two years. As I told a bipartisan congressional committee, I'm here to discuss my professional counterintelligence experience, not my personal life. I intend to follow the same principle in this book.

Whatever my faults, let me also be clear about this: at no point did I violate my oath of office. Throughout, I remained the professional that I was trained to be. I honored my promise — which I made when I was first sworn into the FBI — to "support and defend the Constitution of the United States against all enemies, foreign and domestic." Nothing, least of all my personal political opinions, affected my ability and willingness to carry out the task before me and my colleagues.

My patriotic compass and sense of duty dictated the path that I followed, which was to pursue the facts where they lay — and in the summer of 2016, the trail of evidence had led us to confront a question no law enforcement or intelligence officer would want to contemplate: does the president of the United States pose a national security threat? That was not something I relished doing; what public servant would? But when duty called, I answered. That same devotion to country guided everyone I worked with, and they would have accepted no less in me.

Since leaving the FBI in the late summer of 2018, I have often been dismayed by the small number of thoughtful governmental voices speaking out about the national security and counterintelligence threats that I helped to identify and investigate during my time at the Bureau, which have only grown in severity and prominence since my departure. The conclusion of special counsel Robert Mueller's work, the continued attacks by the administration on the FBI and the intelligence community, and the pattern of Trump's behavior that ultimately led to an impeachment investigation all have strengthened my desire to speak out — and to make the record clear as to why the FBI sought answers to legitimate questions about the Russian connections to first a presidential campaign and then a presidential administration. And clarity on this point is essential, not merely for rehabilitating the reputation of the FBI, but also for shoring up the defenses of America as a whole. Because the threat from Russia did not disappear with Trump's election. If anything, it has grown.

COMPROMISED

If my upbringing impressed on me the strength of our democracy, it also showed me, time and again, how quickly a law-based moral order can collapse if good people do nothing. My counterintelligence work over six presidential administrations — Republican and Democratic — gave me a profound understanding of the perennial work of foreign nations like Russia to undercut the U.S. government.

My life and two decades of service also provided me with experience to understand the unprecedented threat of the Trump administration. Specifically, they gave me an in-depth understanding of how foreign intelligence services manipulate American citizens to serve the interests of other nations rather than our own — something Trump appeared to be doing a lot as he maneuvered his way toward and then into the White House.

Throughout the campaign, the election itself, the lame-duck session afterward, and the period following his inauguration in January 2017, Trump's behavior parroted the Kremlin line. He questioned the value of NATO and America's strategic western European engagement. Not for the last time, he actively advanced the theory that someone other than Russia had interfered with the 2016 election. In response to an interviewer's assertion that Putin was a killer, Trump responded, "There are a lot of killers. You think our country's so innocent?"

Throw in all the financial and intelligence links that showed close and pervasive connections among Trump, a host of his campaign members, and Russia, including Trump's own business deals linked to Putin's involvement, and it would be inconceivable or incompetent to suspect nothing. It begged the question "What is his motivation?"

Even before he became president, Trump said and did things that gave the Russian intelligence services the means by which to coerce him — either subtly or explicitly — into taking actions that would benefit their country rather than his. The moment Trump said publicly, "I have no business dealings with Russia," he knew he was lying, Putin knew he was lying, and the FBI had reason to believe he was lying. But American citizens didn't know that. The then-presidential candidate's public denial of his business dealings in Russia signaled to Putin that Trump was more interested in maintaining his personal financial interests than in telling the truth to the American people, and that he needed Putin's complicity to maintain the lie. To use an intelligence term that you will be seeing a lot in this book, in this moment Trump became *compromised.*

Trump's compromising behavior did not begin or end with the lie about his business interests in Russia. The list was long and alarming. Among other things, between that moment and his inauguration in January 2017, he encouraged Russia to hack Hillary Clinton's email; according to his personal attorney Michael Cohen, he clandestinely paid women with whom he had had affairs for their silence; and people connected to his campaign coordinated with WikiLeaks to optimize the release of email stolen by the Russians to help his campaign. All of these actions made Trump vulnerable to coercion by Russia, and now he was behaving in a way that suggested he was indeed being manipulated by our adversary. The dilemma for us was, what was the Bureau going to do about it?

The gravity of that question hung over the Counterintelligence Division conference room as my colleagues and I argued that afternoon in early 2017. Everything was on the table, from the constitutionality of opening a case on the chief executive down to the tactical nitty-gritty of just what an investigation would or could accomplish.

Occasionally I glanced out the window at the view of Main Justice, what we called the headquarters of DOJ, the parent department of the FBI and other federal law enforcement agencies. Looking farther down Pennsylvania Avenue to the east, I could see the U.S. Capitol dome, the enduring seat of our representative democracy and the envy of nations around the world. I knew that if I peered out the window to the west, I would see the White House, the new residence of the man whose frightening counterintelligence vulnerabilities we were discussing at that very moment.

The tense debate in the conference room stretched for hours, long into the afternoon. Nobody in that room sought the thorny position we were in. No one hungered to investigate our newly elected president. But Trump's behavior had made this debate necessary.

The discussion at times grew heated as patience waned and tempers grew short. It seemed as though the air in the room had thinned, as though we were laboring uphill, struggling to keep up the pace while not becoming overwhelmed by the perilous position we found ourselves in. I felt exhausted. All of us were grappling with the questions before us, but also with the burden of responsibility we felt to the American people. It was on their behalf that we were considering investigating a man who had been elected president but who also appeared beholden to foreign interests, untethered to loyalty and duty to country, and willing to pervert the truth for his own ends.

We did not open the investigation that day. It would be several more months before we did so. But in that moment we first came to the agreement that the president of the United States appeared to be under the influence of, or at risk of falling under the influence of, a foreign government. Trump's self-interest had made him vulnerable — and with him, the American people. Where this would lead us, as a republic, no one in the room could know that day. We all knew that Trump's pattern of behavior would continue — and that the counterintelligence threat he posed to our country would grow — unless the truth came out. But our job was not to expose it. Our job was to investigate as covertly as possible to gather the facts for those who would decide what to do with the information we generated. Whether that responsibility would fall to the courts, Congress, or the American people, only time would tell.

PART I

1

Ghost Stories

FROM THE DRIVER'S seat of my white Dodge Intrepid, I could see the loading dock and the back door of a local bank in Cambridge, Massachusetts. On a frosty Tuesday night in January 2001, I sat crammed in the sedan with four other agents, parked in a narrow alley that emptied into Harvard Square, headlights off and engine running. Our battery chargers glowed softly orange in the dark as we topped off power to our phones and cameras and radios, juicing them up one by one using the single cigarette lighter. We wore "soft clothes"—nondescript street clothes, jeans, jackets with lots of pockets, and low boots—so we wouldn't stand out on the street. Backpacks and black Pelican cases containing delicate equipment filled the trunk, our laps, and every other available space in the car. As we waited, I fiddled with the heat, trying to find the sweet spot that would ward off the cold without fogging the windows.

We had carefully chosen the timing of the operation. During the day, tourists and students and buskers packed Harvard Square. We had cased the bank, driving and walking by at all hours, noting when the students emptied out of the bars and restaurants and the foot traffic to the nearby apartments was light. After weeks of spot surveillance, we chose a Tuesday night as the time when our five-member crew was least likely to be seen on its mission to sneak into this bank branch and crack a safety deposit box deep inside an underground vault, which we suspected contained evidence in the biggest spy case of the new century.

The minutes ticked by. I periodically glanced in the rearview mirror. As expected, the alley was quiet and still.

Sometime before midnight, the cleaning crew filed out of the back of the bank, loaded their vacuum, mops, and buckets into a truck, and drove away. We waited a few minutes longer, to make sure they didn't return for a forgotten item. Then we gave the signal to one of the bank's managers, who had been sitting in a car in front of us.

This was the first time I had broken into a bank, so it helped that it was an inside job. A few weeks earlier, my partner, Carl, had approached the manager with a highly sensitive search warrant giving us authority for what we were about to do. A few wary conversations and a meeting with the bank's attorneys confirmed the legality of what we were doing. It helped that Carl was a gruff New Englander from Foxborough who had cut his teeth on Soviet counterintelligence work in New York. His terse and direct manner, along with a carefully groomed white beard and hair, evoked a South Boston Ernest Hemingway. The manager became a reluctant participant, harboring a mix of curiosity and a commonsense desire to have nothing to do with it. Tonight he sat listening to his car radio as he waited with us for the cleaners to leave. Behind us, a locksmith's truck made up the third vehicle in our convoy, carrying a quintessential older Boston Irishman who still had a noticeable brogue.

The manager got out of his car and walked to the loading dock. We knew from scouting the inside of the bank on prior surveillances that the safety deposit boxes were in a vault at the bank's left rear, down a half flight of stairs. The sightlines weren't in our favor — the bank's wide front windows looked out on Massachusetts Avenue, with a clear view back toward the stairwell. But a small alcove beside the vault door was large enough to get everyone and our equipment down onto the lower landing and out of sight of the Harvard Square traffic.

From the car we watched the manager unlock the bank's back door, disappear inside to make sure the alarms and cameras had been deactivated, and come back out. We looked up and down the quiet back street: no pedestrians. We climbed out of the sedan into the cold January air, distributed all the equipment cases and backpacks among the five of us, and walked quickly up the loading dock and through the back door, with the manager leading the way.

The bank was quiet and still. The branch kept the lights on all night for security, and the overhead glare added to a sense of exposure. We slipped past empty administrative offices toward the front, edging close to the high-ceilinged lobby area, mindful of the short but nakedly open gap between us and the stairs descending to the vault. With a glance to the street, we crossed the open area and took the stairs down to the alcove and its unobserved safety.

Outside the vault, a file cabinet held the safety deposit box access cards, which cataloged who went into the vault to view their boxes and when. We were looking for a box rented by Don Heathfield and Tracey Lee Ann Foley. A middle-aged Canadian couple, they lived on Leonard Avenue in mid-Cambridge, and Don attended Harvard's Kennedy School of Government. They had rented the box shortly after their arrival in Cambridge, and Ann had accessed the box a few times. While we worked at the bank, the couple was at home under observation, FBI personnel from the Special Surveillance Group, or SSG, unobtrusively watching from cars parked along every exit path from their apartment building.

The manager let us into the vault, where row upon row of multisized machined metal boxes showed the paint colors and solid craftsmanship of a bygone era. Each safety deposit box required two keys to open. The bank held one key, which we had. To ensure privacy, the customer held the second key, which we didn't have. Our locksmith searched through his collection of uncut keys (called "blanks") for a match, found one, and began to do his painstaking magic, slowly filing down the contours of the blank to match the tumblers inside the cylinder.

While we waited for the locksmith, a member of the team named George leaned over to me. *Remember, don't touch anything—I mean anything —unless you ask me and get my permission,* he said sternly. George was a kind, extraordinarily fit agent who controlled his workspace while out on operations like a drill sergeant. He and another member of the team, PJ, were from a group within the FBI laboratory known as "Flaps and Seals," an old name that stuck through numerous official changes. The name derived from their mission: opening a sealed envelope, cracking open a shipping container, bypassing the alarm system in a locked home, all without leaving a trace. Complicating their job, their adversaries in espionage were trained in detecting intrusion in ways large and small, from memorizing precise placement of papers on a desk to placing a nearly invisible hair on the lip of an unopened dresser drawer, from which the hair would fall when the drawer was opened. I didn't fully appreciate their work until I watched one of them on a later search, long, slender tweezers in hand, fastidiously re-placing dust bunnies at the rear of a computer that we had just searched. A careless motion, a misstep, an inattentive nudge of an object, could ruin an entire mission.

Some people find it easiest just to keep their hands in their pockets, George said.

Our Irish locksmith finished cutting the blank. After he swept up the fine metal shavings from the floor and lip of the box ledge, we turned both keys and eased open the outer door. With the gentleness of a parent moving a sleeping infant from a crib, George gingerly slid the long, slim box from its metal drawer. Then he turned and lowered the box softly onto a lint-free pad he had put on the floor. After a few more minutes of delicate work, he smoothly lifted the front of the hinged top of the box. We peered inside.

At first the contents seemed disappointing. Flaps and Seals photo-graphed the box's interior and carefully lifted out what was inside. There was a certified copy of Don's Canadian birth certificate, some other family records, and a wad of cash. And then, at the very bottom, we found a small stack of 35mm black-and-white photo negative strips, about 20 frames or so. Holding them up to the light, we saw something we hardly expected: demure images of a young woman posing in a forest.

In our age of digital photography, many younger people may never have seen or held a film negative. If you have, you know it's not easy seeing detail in the thumbnail of the frame, because it's small and the lighting is reversed

—what appears light is dark, and what's dark is light. We didn't have the luxury of time in the bank, so we photographed the negatives to investigate later. Then Flaps and Seals delicately placed everything back inside the box exactly as before. They returned the box to its drawer and began the process of erasing our presence in the vault. We retraced our steps out into the frigid air of Harvard Square.

Back at our offices in downtown Boston, we studied the photos for a "tell" about their true history. Perhaps the woman in the photos wasn't Ann —it was hard to tell, as her face was artfully obscured. But why else would Don and Ann have kept the negatives unless it was her? We scoured the background for something that might have been overlooked, an object left in the camera frame or a building in the distance. We studied the trees and their leaves, their bark, anything that would give away something about the location where these photos had been taken, but to no avail.

And then we found what we were looking for. The clue wasn't the features of the subject or her surroundings but the negative itself. When I examined the images, I noticed something strange. Every strip of film negative had notches snipped out at regular intervals along the bottom. That is, all except one. That one showed a perfectly preserved 20-something woman at the beginning of a life left behind. In that photograph, the notch would have destroyed the edge of that perfect photograph, and so the border had been left untouched to preserve the photo intact.

Printed in tiny letters on the uncut border of that single frame, we saw a word we didn't recognize: *TACMA.* It was the spot on the border where in the U.S. you would see the word *Kodak,* identifying the make of the film. None of us had heard of Tacma. Research didn't turn up anything either.

Then one of us realized that in Cyrillic, a *C* transliterates as an *S,* rendering the word *TASMA.*

Tasma was one of the oldest camera film companies in the former Soviet Union.

After millions of rubles of training and years spent carefully infiltrating the U.S., there on the negative, sitting in a subterranean vault just miles from where the United States was born, Don and Ann had left us the very thing we had broken into the bank to find: evidence linking them to their secret motherland, Russia.

THE LONG GAME

My counterintelligence career began with Russians, and it ended with Russians. The uncloaking of Donald H. Heathfield and Tracey Lee Ann Foley, two Russian intelligence officers who presented themselves to the world and their own children as an unassuming Canadian scholar married to a real estate agent, was but one of the many times in my service at the Bureau that I helped to uncover Russian intelligence operations within the United States. What we did in this case, later known as Ghost Stories, represents the kind of solid counterintelligence work that is a core part of the FBI's national security responsibilities. It was a memorable case, to be sure, but only a microchapter in a long and storied geopolitical struggle between these two world powers, the United States and Russia.

A brief background discussion of intelligence and counterintelligence is necessary to understand the context of this book. While the wary dance between our two nations plays out in the open and in the headlines, it also unfolds out of sight and under cover. In this century-old game of cat and mouse, each nation attempts to tap the secrets of the other while at the same time trying to foil the efforts of the other. Counterintelligence work against countries like Russia and, more recently, China never finishes; it endures without beginning or end. In this clandestine twilight of intelligence and espionage, no allegiances are above suspicion, no information is ever completely certain. Identities shift and change. What appears obvious from one perspective shatters and falls away when viewed from a different angle. Everything is built on sand. Intelligence officers change and special agents come and go, but the game goes on.

For the most part, intelligence work remains out of sight. Unknown to the public, counterintelligence investigations are going on around them in scores of cities across America each and every day — until that silence is broken with news of a spy being arrested or with the exposure of a ring of foreign agents. Or, more recently, when a hostile nation engages in a multipronged attack on our presidential elections, thrusting discussion of the ins and outs of the hidden world of counterintelligence into the public's imagination.

Understanding the FBI's work in counterintelligence — known as "CI" within the U.S. intelligence community — means understanding intelligence. Intelligence is tricky to define. Ask a hundred professionals to explain it and you'll get a different answer from each person. My definition of intelligence is this: clandestine activities conducted by or on behalf of a country to convey a strategic advantage to its nation.

Counterintelligence is the effort to thwart an adversary's secret intelligence work. It has proactive aspects, such as protecting the targets of an intelligence service before they come under attack. It also has reactive elements, such as blocking or neutralizing what foreign intelligence services and their domestic assets are already doing. (Counterespionage — related but different — is catching U.S. spies who provide information to a foreign nation, usually, but not always, with the goal of trying to prosecute them.)

There is a tension between gathering counterintelligence information and stopping the harm being done by hostile intelligence activity. Similarly, prosecuting a case in court is almost always at odds with protecting sensitive intelligence sources and methods. The rules of evidence and adversarial defense work simply do not lend themselves to keeping secrets secret.

But criminal prosecution is only one of many tools for combating an intelligence threat. Pure counterintelligence work doesn't require agents to gather proof beyond a reasonable doubt that a crime took place; it just requires them to countervail the hostile intelligence activity, quietly and hopefully unnoticed. Rather than prosecuting those involved in the hostile activity, counterintelligence agents can try to turn defense into offense. We can introduce false information to lure an enemy into believing something that isn't true. Or we might confront our U.S. subjects, finding a way to get them to come clean and doubling them back against their handlers — the foreign intelligence officers who were running them — to learn more about the foreign intelligence service, what it wants, and the tradecraft it is using to handle its sources.

At the end of the day, though, what separates an FBI counterintelligence agent from most of the U.S. intelligence community is the possibility of his or her work leading to prosecution of suspected wrongdoers in U.S. courts. The evidence developed during our counterintelligence investigations may form the basis of a legal action in which it must be proved beyond a reasonable doubt that a crime has taken place. And the evidentiary rules of a U.S. court must be followed in establishing that laws were broken. Unlike the Central Intelligence Agency (CIA), which conducts intelligence activity but has no direct law enforcement role, the FBI is a Janus-like entity, with both intelligence and law enforcement roles.

In short, we want to stop spies like Don and Ann, but we also might have to bring them to a judge in handcuffs or prove to a jury that they violated United States law. That requires detailing with precision, in an open way consistent with our system of justice, what we know and why we know it, subject to the rigors of adversarial defense questioning. In many ways it's the antithesis of the clandestine nature of most intelligence work. Regardless of what we do, it must accord with U.S. law, a requirement absent from much of the intelligence community's work overseas.

The FBI has the lead role in conducting counterintelligence within the U.S., although it is not the only federal agency that does this important work. While all member agencies of the intelligence community have critical counterintelligence functions, the CIA and the National Security Agency (NSA) play an outsized role alongside the FBI in performing the bulk of the nation's counterintelligence work. All three of these agencies have an especially important role to play in counterintelligence, and often do it hand in hand. But we sometimes have different goals. For example, at its core the CIA collects intelligence; as an organization, information is the coin of the realm. While the FBI and CIA might both be aware of a spy in our midst, the CIA has always been interested in identifying everything we can about the operation: who the spy met; where the spy met his or her handlers; how they communicated; what the spy was tasked to provide. All those pieces of information provide a unique window into the hidden world of a foreign nation's needs and vulnerabilities.

The FBI collects intelligence too, but the Bureau places a comparative priority on mitigating foreign intelligence efforts. So while watching a spy might be giving us a tremendous window into the clandestine world of a foreign adversary, the spy also likely will be hurting U.S. interests on a regular basis, which we are tasked with stopping. This isn't to say that the agencies don't understand each side of the equation, or that one side is more valid than the other — it's just that there is sometimes a divergence of goals, a disagreement in which the CIA tends to fall on the side of continued intelligence collection or quiet neutralization while the FBI wants not just to stop the bleeding but to arrest and support the prosecution of a spy.

The FBI and CIA also use different terms, which adds to the confusion. First, the word *agent*. In the FBI, *agent* generally refers to an FBI special agent, whereas individuals we recruit to work with us are "confidential human sources" — CHSes — or simply "sources." The CIA and most intelligence services use *agent* to refer to a human source, not an FBI agent. The CIA complements to FBI special agents are case officers, which the agency refers to as "intelligence officers," or just "officers" for short.

FBI agents and CIA intelligence officers make up a small component of the people involved in gathering human intelligence. The bulk of actual collection is done — using the FBI's terms — by sources recruited by agents. Sources fulfill a variety of roles. What popularly comes to mind in public imagination when thinking of a source is a spy, a fully recruited source within the intelligence community, aware of his or her recruitment, responsive to tasking, and handled professionally by an intelligence service. In our time, that's someone like Aldrich Ames or Robert Hanssen — arguably the most damaging, but also the most rare. But there are a slew of other types, like spotters, who identify others who might be sympathetic to working for us; access agents, who serve to invite us to places or introduce us to others; and agents of influence, who advance a line of thought helpful to us. Their roles are as diverse and arcane as the practice of intelligence itself. One bond unites them: the secrecy of their work for us.

FBI agents and intelligence officers around the world approach people in the same way: what value could this person provide, and what motivations or vulnerabilities does he have that we can use to get him to work with us? FBI sources are motivated by a variety of reasons, ranging from altruistic to cooperative to coercive. Rarely is there one simple motive. I ran sources who worked with us because they were patriotic Americans. Others needed the money we paid them. For still others, our work made them feel valuable, that not only were they making a difference to America but they were doing something exciting and interesting that made them exciting and interesting. I've spoken with foreign officials who wanted their children to attend American universities. Others wanted to escape scrutiny and repression from their intrusive government back home.

The FBI has enormous counterintelligence capacity, but there's a limit to how much we can do, and adapting nimbly to ever-changing threats and priorities is central to our mission. Traditional counterintelligence against so-called symmetric threat activity — symmetric because Russia and its client states, our Cold War adversaries, were similar in structure and operations to ours — includes ferreting out and watching the intelligence officers assigned to the various embassies and consulates throughout the U.S., monitoring visitors with intelligence affiliations, and from that, learning their sources and methods and what they're trying to collect. Much of this activity is conducted with an eye toward going on the offense, such as recruiting the very intelligence officers we're investigating or launching a CI operation, like sending in false volunteers or reversing the allegiance of their sources by turning them into double agents.

In contrast, asymmetric threats are all those activities outside that definition. Things like visiting academics meeting a nuclear scientist at a Department of Energy laboratory, then getting debriefed by their government once they return from the U.S. Or a Russian contractor sitting in St. Petersburg targeting interest groups on social media to unwittingly sway public opinion in a U.S. election.

If done well, offensive counterintelligence is effective and devious. A double agent who feeds bogus military research to a foreign military intelligence officer can cause that adversary to waste years and millions of dollars pursuing dead-end technology. Creating credible suspicion of a mole can cause an opposing service to believe they have a spy in their midst, grinding their operations and personnel to a halt as they search for the nonexistent spy. And it extends into what the Russians call "active measures," using skewed or false information to achieve political or social objectives, something I would learn later in my time in Boston. At the same time we're trying to combat the threats posed by our adversaries, they're doing the same thing to us. It's an enduring game played in the shadows.

Counterintelligence work is murky, and hardly ever ends with clear conclusions. Rather than the criminal legal standards of probable cause and proof beyond a reasonable doubt, counterintelligence work almost always operates within a much less definite state of belief. Good counterintelligence officers rarely feel that they know much with certainty. It also means that sometimes they know enough to be confident that something happened but do not have enough unclassified or admissible evidence to prosecute it criminally. That's true of most FBI investigative disciplines, and it's particularly consequential in counterintelligence. "We can't prove this beyond a reasonable doubt in a court of law" often does not mean "This didn't happen."

One of the harshest realities that special agents must accept is learning to live with the knowledge that someone's misdeed may not be punished. This is a bitter pill to swallow, as every FBI agent I know is motivated by a profound desire to pursue justice. In this world, bad people sometimes get away with bad things. Some particularly thoughtful senior agents in Boston tried to convey that hard fact, but I wouldn't feel the full weight of that lesson until the end of my career. During the earlier years, I had the luxury of focusing on our victories — one of the biggest of which was the one we scored in the case of the sleeper agents in Cambridge, Massachusetts.

DEAD DOUBLES

Even within the Bureau, I've never told the story of how we unmasked Don and Ann, whose real names were Andrey Bezrukov and Elena Vavilova. They were Russian "illegals," sleeper agents without any official or diplomatic cover, even without their true Russian identities. They were embedded in American society to quietly gather intelligence, build connections with academics and policymakers, and report back to Moscow. For Andrey and Elena, their adversary was the "Main Enemy," which Russian intelligence services have called the United States since World War II, and they were sent to live and work in the belly of the beast.

The U.S. became the Main Enemy after Germany's defeat and the disintegration of the uneasy wartime alliance between the United States and the Soviet Union. The two countries found themselves on opposite sides of a rivalry that would emerge in the form of the Cold War. Just as NATO and Warsaw Pact forces lined up against each other in Europe, their intelligence services did as well, with no priority greater than planting spies in the other nation. While the collapse of the Soviet Union in 1991 brought about political and economic reforms, the Russians' intelligence priorities remained unchanged and their aggressive targeting of the U.S. continued unabated. While the U.S. congratulated itself on bringing down the Soviet Union and turned its attention to the new challenges of leadership in what it perceived as a newly unipolar world, the Russian superpower appeared to have gone into sleep mode. But it hadn't; it was merely biding its time and quietly keeping its spy games going.

While their discovery has led Andrey and Elena and their brethren to be viewed as hacks by some in the American public, the reality is that they were two of the most highly valued agents of one of the best, most professional intelligence services in the world. While they were sloppy at times, we might not have caught on to them by their mistakes alone. (Our attention was drawn to them for reasons that are classified and that I can't discuss.) Even after the collapse of the notorious KGB, the Russian foreign intelligence service, rebranded as the SVR, maintained high standards and kept up an expansive program of intelligence operations that would have surprised Americans at the time with its aggressiveness and sophistication. Andrey and Elena and other illegals like them were the crème de la crème of this elite spy force. Sometimes I marvel that we were able to catch them at all — and I worry about how many other illegals we haven't caught yet.

I don't know what caused the lapse in Andrey and Elena's nearly impeccable tradecraft. Why hadn't they trimmed that one Tasma frame, as they had so assiduously with the rest of the negatives? Given the care they displayed throughout their professional lives, I doubt it was sloppiness alone. Perhaps it was vanity, a strong and stubborn urge to forever capture their unblemished youth. Or perhaps it was to preserve something that predated their professional life, to dispel a premonition of an untethered future rife with sacrifice and loss and fear of losing a life left behind.

Their mistake may also have come from a false sense of being invulnerable because of the layers upon layers of deceit that cloaked their true identities. The intelligence term of art used in this case is "dead doubles." To their neighbors and Andrey's classmates, Donald Heathfield and Tracey Lee Ann Foley were a Canadian couple who had moved to Cambridge so that Don could attend Harvard. Those identities were illusions, carefully constructed over decades and recently fictionalized in the FX television spy series *The Americans.*

Don and Ann — or Andrey and Elena — weren't from Canada, but their names were. Years ago, most likely in the 1970s, someone in Directorate S at KGB headquarters — known as Center — contacted a Russian officer who probably worked undercover as a regular embassy employee in Ottawa. Directorate S, which was responsible for illegals, would have sent the order to the officer, a so-called Line N officer, to find a list of Canadian children who had died in the early 1960s. Officers in the KGB, and now the SVR, are assigned to "lines" based on the focus of their intelligence collection and activities. For example, Line ER handles economic intelligence. Line KR is responsible for counterintelligence, arguably the most significant to the FBI for their role in recruiting and handling penetrations of the U.S. intelligence community. Line N refers to officers supporting illegals.

The Line N officer in Ottawa who received the directive would have scoured cemeteries or searched public records for names of children who had died at an early age in the sixties. Then an officer — either the original Line N officer or another one, to keep information about the illegals compartmentalized — would have followed this initial intelligence gathering with visits to the towns where each deceased child had been born and lived. They might have snapped pictures of the neighborhood where they lived and the area schools. They might have jotted notes about the location of the library, the local theater, and the closest grocery store. All the things familiar to someone who had actually grown up there. All the things that would become parts of Andrey's and Elena's assumed identities.

Center would have then turned these names into so-called dead doubles. That means pairs of KGB agents in training would be assigned the identities of two long-dead children whom they matched in age and gender. In many cases death records, particularly those before the explosion in governmental collection of biometric data, are not tied to central government records. In nations where there is no central identification requirement — which includes most of the West — there is no logical place where such data would come together, especially for the very young, who would never have registered for social services, health insurance, or a passport. They would, however, have received a birth certificate, such as the one we found for the real Donald Heathfield in the Cambridge bank's safety deposit box.

After they assumed their stolen identities, the illegals set about building their fake lives, a process that often took decades. The two KGB officers who had been paired in training were usually then paired in marriage and began the slow process of infiltrating the United States using the names of the deceased children and their fabricated backstories. They had pledged their lives to Russia, and their assignments were for life.

Andrey and Elena created a life together, shedding their Russian identities and assuming Canadian ones, then moving from a staging hop in Europe into Canada. The Soviet youth who attended college in Siberia was now a Canadian thirtysomething named Donald. He earned a degree from a Canadian college, then a master's degree in business administration in

Paris. From there he applied to and was accepted into the Kennedy School of Government at Harvard. He and Ann found an apartment on Leonard Avenue in mid-Cambridge. They settled in, and their Canadian-born children, Tim and Alex, began attending L'Ecole Bilingue, a private French-English school.

As they built their fictitious life, the world was changing. Proxy wars began and ended around the globe. With support from the West, the Solidarity movement gained strength and power in Poland. The Berlin Wall fell. The Soviet Union collapsed and rebellious satellite republics gained their independence. The mighty USSR shriveled to just Russia once again. The Cold War ended, and for a brief time relations between the United States and Russia thawed. In the post-Soviet era, the command economy of communism collapsed. President Boris Yeltsin began selling off state enterprises that had previously been owned by the government. The era of oligarchy in Russia began, with ultrarich businessmen — many with ties to organized crime — amassing vast wealth and power. The KGB turned into the SVR. A low-level former KGB agent named Vladimir Putin steadily rose through the ranks for government. On August 9, 1999, Yeltsin named him prime minister.

Through it all, the small cadre of elite sleeper agents we would later identify, including Andrey and Elena, continued their careful work of observing life in the United States, gathering tidbits of information, building friendships and trust with new acquaintances, all the while reporting the information back to Center. Until, unbeknown to them, and for reasons that are still classified, they attracted the interest of the FBI.

At the beginning of the investigation, my partner, Carl, and I would meet every week at a pizza parlor near Harvard Square. Same time, same place every week. Sometimes we'd grab a slice and a soda to go; other times we'd just get to work. Around 6:45 p.m., I'd climb into my Dodge Intrepid and Carl would stroll along the sidewalk nearby, for all appearances another Boston worker returning home for the night. I'd adjust my shortwave radio to the right frequency. Like clockwork, I would hear a Morse Code signal chirping away from deep inside Russia, mirroring a radio inside a closed-off room, chosen because it was on the opposite side of Andrey and Elena's apartment from the bedroom of their two young sons. As Russia's geopolitical realities and the technological landscape evolved, so did their tradecraft; after transmitting from near Moscow, broadcasts shifted to Lourdes, Cuba, and then back to Moscow, before finally giving way to Internet-based forms of covert communication. Despite the advance of technology, Russian trust in tried-and-true methods never went away. Andrey and Elena kept their shortwave radio for backup communications and still had it in their house when we eventually arrested them in 2010.

The broadcast carried a weekly message from Center, in the form of hundreds of five-digit groups of numbers. They were scrambled with a one-time pad, a cryptography method dating to the late 19th century that required sender and recipient of codes to have matching pads printed with identical sequences of randomized numbers. The random numbers, when added to the numeric values of each letter in the message, encrypted the message contents. Those numeric codes were then broadcast, and the recipient with an identical pad decrypted the numbers. After each code was sent, the used encryption page was torn off and destroyed, and the next page of random numbers was used for a new message.

While I listened to the broadcasts from Center, Carl would note which lights in which rooms would be turned on during the coded communications. Every week, Elena, the KGB-trained communicator of the two, wrote down the set of numbers through an earpiece plugged into the shortwave radio. Hundreds of miles away, specialists in the FBI listened and did the same.

We had yet to decipher the broadcasts, and if they were encrypted well, we might never crack them. But we made sure to monitor them without fail, because even the apparently random five-digit Morse Code groups held clues to the illegals' secret lives. An unusually long broadcast might mean something major was brewing. In another puzzling lapse of tradecraft, Andrey and Elena's broadcasts paused when they were away from Cambridge, confirming that the broadcasts were intended for them and no one else. Sometimes they would send coded messages to Center on a special analog

of carbon paper, which left invisible ink underneath an apparently innocuous letter to an accommodation address, a false address somewhere in an
innocent country overseas that appeared to belong to a friend or business
but was secretly maintained and serviced by another Line N officer. But
always, every Thursday, for years, Morse Code messages arrived from the
motherland. And Carl and I were there to monitor them.

We were at the infancy of the investigation, fastidiously building proof
that Don and Ann had a Russia connection, gaining momentum as we
developed probable cause to break into their apartment and their safety
deposit box. We followed them discreetly around town; we snatched their
garbage bags and combed through the trash to identify partial clues from
scraps of receipts and discarded mail. As we identified their phone numbers, we obtained court authority to listen in, and as we monitored them,
we identified yet more ways they were communicating — cell phones, email
addresses. All were added to the rapidly burgeoning stream of information
flowing into our offices at 1 Center Plaza in downtown Boston.

But the real payoff came when we broke into Andrey and Elena's apartment. Months of surveillance had taught us what their neighbors looked
like, how they dressed, who worked the night shift, who owned dogs and
when they walked them. We knew their neighborhood better than our own.
Dressed like a Harvard student and his father (or at least that's how I presented it, an ageist dig to aggravate Carl), we entered their building when we
knew they were not home. But we couldn't just walk into their apartment.
They lived on the middle floor of a triple-decker, the three-story clapboard
apartment buildings built in the early 20th century that are ubiquitous in
the Boston area. These buildings have shared entrances, common stairwells
up to the top floor, and thin walls and ceilings. Many of the other residents
thought they knew Don and Ann, but they didn't know us. To make sure
the neighbors didn't make a surprise appearance during a break-in, we had
to know everything about who lived above, below, and around the family.
We monitored their schedules and workplaces, making sure they wouldn't
be home to see suspicious activity in their neighbors' window and call the
police. What's more, once inside, we'd be able to work fast: thanks to our

nighttime shortwave broadcast work, we had a pretty good sense of which room they used for their covert communications.

When Carl and I broke into Andrey and Elena's apartment, a Flaps and Seals team came with us to make sure we perfectly covered our tracks and left nothing out of place. The kit that the team traveled with would be the envy of any professional burglar. They also brought cameras. Although digital photography was supplanting film, old habits die hard, and instant photography had its benefits. Every search consumed prodigious amounts of film, with pictures of everything from the location of the feet of a chair on the floor to the alignment of individual files and papers within a filing cabinet.

As we moved from room to room in the apartment, the floors would fill with neat little stacks of Polaroid pictures, aids to recreating the scene exactly as it appeared upon entering as well as for detailed review and analysis after the fact to tease out previously unnoticed investigative information. Hiding the photography from intrusive eyes presented its own problem, for which the Flaps and Seals team carried lightproof sheets, secured if necessary with no-residue gaffer's tape, to prevent flashes from alerting neighbors or passersby.

During our first break-in, all these precautions, and all our other hard work, paid off. We were able to clone the hard drive of a suspicious laptop and diskettes and photograph documents and items that appeared to be potentially incriminating. But it wasn't until we got a look inside their safety deposit box that we finally got direct evidence that we were dealing with Russians.

Illegals like Andrey and Elena represented one of the rarest elements of spycraft, and the most clandestine. A nation's intelligence officers — trained government employees posted abroad — are the clearest and most obvious figures in intelligence work. They are posted under official cover, although they can be either declared or undeclared. Declared officers are what they sound like: their intelligence affiliation is acknowledged to the nation they're stationed in. Far more frequently, intelligence officers are posted under cover, which in turn comes in its own two flavors: official and nonofficial. In the former case, think of an SVR officer claiming to be, say, a congressional attaché for the Russian Ministry of Foreign Affairs, based at the embassy. The latter case — nonofficial — can be someone without any diplomatic accreditation who usually, but not always, maintains his or her true nationality. Officers under nonofficial cover, known as NOCs, enjoy a much greater degree of clandestinity — or secrecy, in layman's terms — than those assigned to an embassy, but lack the most powerful protection afforded an embassy official: diplomatic immunity. If they get caught committing a crime, they're probably going to jail, and only quiet diplomacy or a spy swap will get them back home. The final category involves the deepest cover, the illegals like Andrey and Elena. To the world around them, they aren't diplomats or even Russians. They are just two ordinary civilians going about their lives.

As an illegal, the more you limit activities relating to your past life, the better off you are. Andrey and Elena, who were two young Soviets who barely knew each until their country paired them for a lifetime of espionage, had two kids who were teenagers by the time their parents were arrested. We eavesdropped on their domestic disputes, their birthday parties, the ups and downs that any couple would have over the course of years of marriage. After following their activities closely for many years, we learned a great deal more about them.

Most people have no idea how many people it takes to work cases this complex, and I hate hearing that I was *the* person responsible for anything. The reality is that it requires an extraordinary team, in which everyone plays a small but vital role. Carl led the way in Boston, and I learned from him, building on that experience in my own cases and later passing it to agents on other squads that I supervised in Washington, D.C., and at headquarters.

And we were only the beginning. Over the course of a decade, an additional cast of scores of colleagues around the FBI and across the U.S. intelligence community patiently watched the network, meticulously cataloging their contacts, their tradecraft, gathering precious and unprecedented detail into the deeply hidden world of Russian illegals. While Carl and I tracked

Andrey and Elena over many years, beginning around the new millennium, it took an extraordinary group, proud secret teammates, to arrest them and nine of their colleagues in a sweeping spy ring takedown over a decade after our investigation began.

The extensive surveillance that we conducted over that long period felt intrusive, and it was. In truth, I never got fully comfortable with it, and I'm glad I didn't. Don't get me wrong — these were necessary, legal steps. But they were not ones to be taken lightly. What's more, as law enforcement officers who have sworn an oath "to support and defend the Constitution of the United States," I and my fellow FBI agents were bound by clear rules about what we could and couldn't surveil and how we could go about doing it.

Throughout the FBI's investigative guidelines runs the idea of using the "least intrusive method" to achieve what needs to be done to protect the U.S. It's an idea tightly tethered to the role and limits of government in our private lives, honed through lessons of abuse and the hopefully perfecting influence of oversight. Moreover, it's fundamental to who we are as a nation. I always watched new special agents carefully when they experienced their first truly invasive surveillance. In my mind, that ought to feel uncomfortable; it should be cause for introspection and quiet, powerful discussions at work and during informal conversations over a couple of beers. *I really feel . . . I don't know, almost inappropriately voyeuristic about this. Do we need to be doing it?* Those are exactly the concerns I hoped to hear from my agents and to always maintain in myself. If they didn't, if I didn't, then it was time to step back, recalibrate, and rebalance what we were investigating against what we were defending.

Years later, as I was investigating another, much more disturbing counterintelligence case involving Russia, I would find myself interrogating my own investigative decisions even more than in this moment. These cases are like that: the higher the stakes, the more you're aware of your responsibility as an FBI agent tasked with rooting out foreign intelligence threats in service of and within the bounds of U.S. laws. But in that later case, as in the case of Andrey and Elena, intrusive surveillance was done lawfully, and was necessary to protect our nation.

Moreover, surveillance is part of the unwritten rules of the game. Officers who serve a foreign intelligence service knowingly accept the risk and the consequences of spying on U.S. soil. Andrey and Elena knew from their training, during which they had performed at the top of their class to earn the right to deploy to the United States, that the Main Enemy had the most advanced and competent counterintelligence environment in the world and might identify and surveil them. Andrey and Elena and their illegal brethren, just like CIA officers overseas, knew that the cost of doing intelligence business was that they might fall under host nation surveillance. It's just one of the cold rules of the road.

2

Cake and Handcuffs

WE KNEW THAT Andrey and Elena were too well trained to say anything truly interesting over the phone. Rather, they were likely to open up about their illegal intelligence activities only in places where they thought they were alone and could speak privately. Places like their car.

Installing a bug takes about 40 seconds on a television show, but in reality the well-planned, long-term placement of a microphone in a vehicle is an elaborate, complicated procedure. You have to understand the layout of the car, its electrical system, locations in the interior where you can hide a microphone that will actually pick up the sound of a quiet conversation. Then you have to get hold of the car, with both enough warning and enough time to do the work of installation. In Andrey and Elena's case, listening to their every move, we discovered that their car needed maintenance, which gave us a logistical leg up for the elaborate planning that followed.

The plan was a bit of a coordinated dance, because it required three teams working in tandem: the technical team that would install the equipment, Flaps and Seals to clear the car before and after the technicians did their work, and a surveillance team to ensure that Elena didn't unexpectedly return while the maintenance was being done on the car. Flaps and Seals, flown into town just for this job, would clear and open the car, the tech agents would work their magic, and with any luck we'd be free for a late lunch.

The reason we went to such great lengths to know what Andrey and

Elena were doing was that their work clearly was important to Moscow. They weren't running agents, an intelligence activity that carries a significant amount of risk because of how much it exposes the handling officer. Instead, Andrey and Elena provided two golden items for the Russians. The first was the ultimate backup system in the event the U.S. threw out or otherwise neutralized all the Russian intelligence officers under official cover in the country. The second, more insidious benefit was the constant flow of assessment information they sent back to Center. As Andrey went through life as Don, the Kennedy School student, he rubbed elbows with students from around the U.S. and the world. Many were connected to their respective governments or destined for powerful positions in business. Andrey would note their personalities, their likes and dislikes, their quirks. Did they have a hobby, like cycling? Were they prone to drinking too much and talking too much when they did? All of it was quietly noted, shared with Elena, and secretly sent back to Moscow, lining the Center's files with hundreds upon hundreds of people who might be of use to the Russians right then and there or who might be promoted up through the ranks for many years, until they were suddenly in a prime spot for the Russians to target them.

The couple's value was so high to the FBI, in short, because it was so high to Moscow. Andrey, Elena, and their illegal brethren represented the most highly evolved art of clandestine work — their painstakingly crafted identities, decades and decades of their lives, their jobs, their communications, their funding, were all due to the work of scores and scores of supporting SVR personnel. The resources spent to get them into position were matched only by the care expended to keep them there.

Clandestine operations require careful effort to hide the role of the intelligence agency in the activity. A shadowy hand guides all intelligence work; secrecy is the veil that shields that hand from view, either obscuring the origins and intentions of that hand or hiding it altogether. A variety of terms describe methods for hiding that activity, such as "cover for action" and "plausible deniability."

But intelligence agencies don't simply conceal an intelligence activity. They engage in a formal, deliberate process to make their action appear to be something else, which is both innocuous and more logical to the setting —for instance, planning a meeting at a playground frequented by the children of both the intelligence officer and her recruited agent. These sorts of deceptive tactics are part of why the Russians in particular are so effective—and so dangerous. Especially when they're aimed at someone in a position of power or with access to U.S. secrets, such as a government employee.

As I saw time and again, especially while investigating another spate of Russian intelligence activity against the United States, a competent intelligence service won't throw an intelligence officer directly at a U.S. government employee on the street. Instead, that service might identify a conference that the U.S. employee is attending, find an academic attending the same conference who isn't an intelligence officer but who has agreed to serve as a friendly contact in the past, and use that academic as an access agent to make an initial approach to the employee. Another scenario: sending a young female gun enthusiast from Russia to study in the United States, ingratiate herself with gun lobbyists and the National Rifle Association (NRA), rise through Washington social and political circles, and eventually meet the future president's son. As former Trump campaign manager Paul Manafort said, "It's not like these people wear badges that say, 'I'm a Russian intelligence officer.'"

Some of this may be self-evident—obviously, spying is secret—but it has subtle yet profound impacts on the way counterintelligence professionals perceive the world. FBI counterintelligence agents and CIA intelligence officers alike know that intelligence activity can occur without being noticed or identified, but also that an apparent absence of intelligence activity doesn't mean there's nothing there. Foreign intelligence services aren't only trying to be secret about their activities; they're actively working to hide what they're doing with a formal, highly trained, and well-funded infrastructure.

Much later, when I was investigating Russian meddling in the 2016 presidential election, it bothered me greatly that we couldn't pin down the extent of Russian interference. But it also didn't surprise me. A lack of clarity about what you've uncovered is a common and direct result of clandestine intelligence work.

So when do you stop looking for foreign intelligence activity if you suspect it but can't find proof of it? The unsettling answer is that you frequently don't. Sometimes you *can* prove a negative and successfully determine that an intelligence adversary wasn't involved in an event, such as definitively showing that a Department of Defense employee contacted the Russian embassy purely for the innocent purpose of getting a visa for a Moscow vacation. But that's rare. In most cases, counterintelligence work is like the worst novel you've ever read: a never-ending stream of scenes following an intriguing path that ultimately goes nowhere, all loose ends and no satisfying conclusion.

Facts are stubborn things, as John Adams said — but they are also slippery. Which helps to explain why, in our pursuit of evidence, counterintelligence agents, like other intelligence professionals, sometimes find ourselves having to resort to intrusive investigative techniques. These measures, such as the planting of bugs in Andrey and Elena's car (which we ultimately succeeded in doing), are intended to pin down facts before they slip away.

There are, however, legal and ethical lines that the FBI won't cross in the course of its work but that Russia's intelligence services have no compunction about overstepping. The SVR, indeed, frequently employs other, harsher techniques in its service to Russia — methods aimed at manipulating not information but people.

STRONG-ARM TACTICS

One of Andrey and Elena's primary goals on their mission to the United States was to help the Russian government identify American citizens who could help Russia. These two illegals, like their hidden compatriots across

the country, had been tasked with collecting information about the vulnerabilities of select people with whom they came into contact, information
that might be exploited — even years later — by the SVR to recruit, by coercion if necessary, those people into working for them.

Coercion differs from persuasion, which is what we almost always relied
on, using incentives to induce someone to take an action. The FBI and the
CIA have the benefit of selling the best brand in the world: American democracy, economic opportunity, and religious and personal freedom. All
are appealing and morally persuasive, particularly to someone who can't
take them for granted.

That's a tangible benefit of our national character — in intelligence work,
it matters that America is the shining city on the hill. Certainly our nation's
prestige and reputation for freedom and justice mattered more than any
amount of money we could have paid to countless Russian and Chinese and
former Soviet-bloc intelligence officers who chose, at great personal risk, to
work with us. (This is one of the reasons Russia has had to work as hard as
it has for as long as it has to undermine the legitimacy of our democracy.
If the international perception of our government could be reduced to our
being transactionally corrupt, racist, intolerant, and opportunistic grifters,
well, then we would be no different from all those we criticize — and no
more alluring to would-be sources than our competitors in Russia and elsewhere.)

The opposite of positive persuasion is negative coercion: creating a dis-
incentive for someone to act in a certain way. Coercion requires feeling at
risk. Shame is often the most effective leverage — for example, the public
embarrassment heaped upon someone who has lied about his success, or
the shame of a congressman plied with cash and cars to push through favorable legislation.

Coercion is risky, as people tend to react poorly and unpredictably to
fear. I rarely used it, nor did my colleagues at the FBI regularly employ it.
But while *we* don't pursue the tools of negative coercion much, our adversaries certainly do.

The Russians employ coercion so much that they even have a special word for a particularly effective form of it: *kompromat,* an abbreviation of "compromising material." The Russian intelligence services devote considerable effort and expense to obtain *kompromat* on people they wish to manipulate, whether through sexual advances, bribes, or other inducements, as well as the means to capture evidence of it, like extensive bugging of hotel rooms, phone calls, and email. Thanks to what I consider the unethical behavior of our 45th president and the many different forms of coercive leverage that a hostile intelligence service could have over him, this Russian term of art is one with which Americans unfortunately now are obligated to familiarize themselves.

There is no set formula for compromising someone, and its insidious benefit is that there's almost never a need to formally acknowledge it. Someone's small yet deeply shameful fact, almost innocuous to the world around him, might be sufficient — if you can find it — to cause him to turn against his country. What matters isn't so much the shameful fact itself but how the person views it — and thus how much power it wields over him.

That's why Trump raises so many alarms from a counterintelligence perspective. It wasn't only the people around him, his campaign, and his administration being charged and convicted. It wasn't the many alleged indiscretions or suspect business dealings themselves; it was that time and again he sought to cover them up or tamp them down. It wasn't his multiple alleged affairs; it was that evidence showed he had dispatched his fixer to buy the silence of the women involved. It wasn't that he wanted to build a multimillion-dollar Trump Tower Moscow; it was that he misled the public about it over and over, claiming that he had no business deals at the same time that negotiations over the tower continued. At the time he did so, the Trump camp knew it, the Russians knew it, but the American people didn't know it.

As these deceptions piled up, the foundation of Trump's snowballing campaign success became irrevocably tethered to his duplicity. Worse, the parties involved — even at arm's length, even separated by an ocean — had to understand their interdependence. They recognized Trump's need to maintain the facade, and therefore the coercive power behind the deceit. It was a gift to the Russians, and Trump kept giving.

This is how *kompromat* works. There's no formal agreement — no handshake or document. The compromised liar need not be told what to do. It all unspools without anyone's ever having to say a word. Trump's apparent lies — public, sustained, refutable, and damaging if exposed — are an intelligence officer's dream. For that very reason, they are also a counterintelligence officer's nightmare.

When FBI headquarters made the decision to take down the illegals network in July 2010, all that was yet to come. Trump had just completed filming season 10 of *The Apprentice*. Paul Manafort, then a lobbyist fresh off work to successfully install the pro-Russian Viktor Yanukovych as the Ukrainian president in February, was busy receiving $10 million from Russian oligarch Oleg Deripaska, a debt that would follow him to the presidential campaign of 2016.

Like many sensitive counterintelligence investigations, the illegals cases had been closely managed from headquarters. It had been a remarkable run, but by early 2010, the leadership team at headquarters, working with the White House and the intelligence community, decided it was time to move. We were bringing in the illegals.

RAID JACKETS

In the early summer of 2010, I was supervising an espionage squad at the FBI's Washington field office near Judiciary Square. From my fifth-floor office, I had a view of Holy Rosary Church, Georgetown University Law Center, and a homeless shelter. I was in the midst of a series of Chinese espionage cases when my phone rang on a midweek morning. A longtime friend, a special agent from the Boston office, was on the other end.

We need to talk on a secure line, the friend said.

Hold on, I said. I extracted the cryptographic key from the safe and inserted it into my secure desk phone. With the call now encrypted, my friend swore me to silence, then made an offer no one in their right mind would refuse.

Want to fly up to Boston to help arrest Andrey and Elena? he asked me, using the still-classified FBI codenames for the two.

Yes! I said without hesitation. It was a gracious offer. Boston had more than enough agents to make the arrest, and while my familiarity with Andrey and Elena undoubtedly would be helpful, my presence was far from necessary. But my friend knew how much I would want to be there when the big moment arrived.

He explained that the arrests presented serious logistical challenges. Spread across several cities, many of the illegals traveled frequently, and all of them were extremely alert to any changes in their surroundings. The arrest planning had been under way for some time, but had accelerated in recent weeks because several of the illegals in various cities had imminent travel plans. As it turned out, the trigger for the entire operation was Andrey in Boston, who would soon be flying to Europe with his son Tim. It wasn't clear if Andrey had made plans to return.

As analysts at headquarters looked at the scores of moving pieces, they identified an optimal window for arrests, a time frame determined by Andrey's approaching travel overseas: on Sunday, June 27. I flew up to Boston a couple of days before. Surprisingly, an upscale hotel in downtown Boston still had rooms available at the government rate on short notice. The last time I had been at that particular hotel, we had flown up a defector for a false flag operation, asking them to take on their prior identity to fool someone we suspected of spying into believing they were talking to a real intelligence officer from the foreign country for which they had worked and who was there to warn them of an impending emergency. This time there was no subterfuge ahead. The prospect of going in the front door in raid jackets really felt good.

Early in the morning on Saturday, June 26, I walked from the hotel to the field office on Cambridge Street. The shops at Faneuil Hall were still closed, and there was little traffic so early in the day. It was a warm summer morning, overcast, with the temperature at nearly 70 degrees at 7 a.m.

As I walked, I thought about the approaching conclusion to this case I had begun more than a decade earlier. It felt surreal. All cases take root in your soul — they become a living, breathing part of an agent's day-to-day existence. But Ghost Stories was special. It was magnificent in its breadth and depth, representing the best of the Bureau — dozens of cases across the country for over a decade, connected by scores of agents, analysts, and professional staff using the most complex investigative techniques without ever being caught. In little more than 24 hours, secret counterintelligence work that spanned half a generation would be revealed to the world.

I walked with quiet anticipation across the open brick expanse of Government Center to what was then the FBI's Boston field office and met up with Carl, who had retired as an agent a couple of years before but continued as a nonagent employee. There were current case agents whom I was happy to see again. Carl and three of the current case agents were going through their notes, reviewing outlines and display material for the interviews of Andrey and Elena. I sat with Carl in a small conference room, roleplaying a Russian while he walked through his outline again, as he and the others had done countless times in the prior weeks.

Later in the day, the group of agents who would take the Boston illegals into custody assembled in a conference room. Arrest teams get large quickly. In this group we easily had 40. Andrey and Elena got two agents apiece, while their sons, Tim and Alex, got a single agent each, along with social workers, as Alex was still a minor and would need to be handled according to the procedures established to protect children. There were agents assigned to knock on (then break down, if needed) the door and enter first to clear and secure the residence. Eight or more evidence experts and photographers would search the residence following the arrest, along with two computer forensic experts. Flaps and Seals would come along to lend their memories from past covert searches. Command post personnel supported the effort from the field office and coordinated with FBI headquarters in Washington. There were Cambridge Police Department officers to control the scene, supervisors to oversee the entire effort.

And I would be going in the door as a member of the arrest team and then remain behind to help search for evidence of Andrey and Elena's secret tradecraft.

The arrest briefing imparted crucial information, but it also conveyed a reassuring sense of structure and predictability. The communal recitation of the well-known FBI arrest plan format felt a little bit like a religious mass, as if we were fellow worshippers reciting familiar scripture by heart. We heard a description of the subjects, followed by a detailed execution plan, the obligatory reading of the FBI's deadly force policy, reserved radio channels to be used, the nearest trauma center, and contact information for the chain of command and local law enforcement. After questions, everyone made last-minute equipment adjustments and lined up the vehicles in the basement garage that would convey everyone to the arrest in the morning. And then, in the early evening, we dispersed toward home. Everyone except the surveillance teams, whose members, unblinking as owls, watched over Andrey and Elena through the night.

On Sunday morning — arrest day — I put on a light gray suit and black loafers and retraced my route across the deserted brick expanse of Boston City Plaza to the field office. When I arrived, I sensed a palpable energy charging through the office. Even Carl, forever taciturn with a dash of habitual crankiness, had a type of pregame excitement that I had not seen before. He carefully double-checked the inventory of notes, outlines, photographs, and other documents that he had standing ready to use in the interview.

We waited. Some agents went out for a quick bite to eat, while some disappeared to use the bathroom. Others again made last-minute adjustments to equipment. Some just sat quietly in the squad bay area, making small talk to stay relaxed and pass the time. Finally the order to load onto vehicles came and we were ready to roll.

Around midmorning, our convoy of a half-dozen or so sedans and trucks rumbled out of the field office garage toward the staging point in Cambridge, several blocks away from the residence. Andrey and Elena had moved twice since I had left Boston. The most recent move had been only a few weeks earlier, to a mid-Cambridge townhome on Trowbridge Street, a tree-lined street of Victorian houses and apartment buildings a few blocks east of Harvard Yard and a short distance from their first apartment in the U.S. As we headed toward Andrey and Elena's building, arrest teams in Yonkers, New York; Montclair, New Jersey; and Arlington, Virginia, simultaneously streamed toward their targets' homes.

At the staging location, we waited in a line of idling vehicles for the green light from headquarters. We knew what to do. The entire operation had been carefully choreographed, down to the timing of the traffic lights and the distance to the front door. Every variable had been accounted for, every escape and egress controlled. Any arrest depends on getting to the subjects as quickly and safely as possible, with as little noise and advance notice as possible. Ideally, neither Andrey nor Elena would have any warning until the moment they looked into the eyes of their arresting agents.

At midday we got the signal from headquarters. A surge of adrenaline leapt through me. The convoy of sedans and trucks roared into gear and sped toward Andrey and Elena's home. At the same moment the arrest teams in Yonkers, Montclair, and Arlington sprang into motion as well. Across the country, the invisible net that had surrounded the illegals for years was suddenly becoming visible — and was rapidly tightening.

Our vehicles raced through the winding Cambridge streets scant inches apart, trying to time stoplights but using emergency lights to stay together when needed. The plan called for the convoy to turn up the one-way road of the residence, after which Cambridge Police Department vehicles would seal off both ends of the block. By the time the marked police vehicles pulled into place, the entry team would already be knocking at the front door, a battering ram at the ready.

The fast-moving line of cars turned onto Harvard Street for the final leg of the route to Trowbridge Street — but the convoy came to a sudden halt. The lead vehicle radioed back.

Cambridge PD's blocked the road.

A city police officer standing by to assist the operation had inexplicably blocked Trowbridge Street, lights blazing, erroneously thinking it would be helpful to stop traffic to the residence. Murphy's Law in action.

An agent from the lead vehicle jumped out and raced to the officer in his vehicle. *Hey, we need to get down this road,* he explained.

I can't let you through, there's law enforcement activity going on, the officer responded.

We're *the law enforcement activity,* the agent shot back.

I can only imagine the police officer's thoughts as he looked down at the long line of unmarked federal vehicles and realized he was stopping the entire FBI operation in its tracks. He slammed his vehicle into gear and bounced up onto the brick sidewalk in reverse, allowing us to stream through, before pulling forward again after we passed, sealing up the road again. The convoy screeched to a halt in the middle of Trowbridge Street, all the doors sprang open, and dozens of agents bounded out toward the residence, hoping we still had the element of surprise.

Andrey and Elena's home was on the right-hand side of a handsome building with clapboard siding and manicured bushes in front. The entry team was the first through the gate in the wooden fence and up the brick walkway to the porch, followed by the arrest team. By the time I entered, agents had separated Andrey and Elena in different corners of the living room and snapped handcuffs on each before patting them down for weapons, needles, or concealed handcuff keys. The scene in the living room was a portrait of what might otherwise have been normal life, interrupted. Their younger son, Alex, stood beside a table with a partially eaten cake and champagne glasses on top. We had burst through the door in the middle of Tim's 20th birthday celebration.

As I surveyed the living room, the surreal tableau had an odd sense of incongruity, an upended family birthday celebration collapsed into a law enforcement action, snapping a cold, detached frame around what otherwise would have been a scene recognizable to any family in America. I knew almost every face I saw — Carl, the arresting agents, Andrey and Elena, and Alex, the Bureau supervisor — but in very different contexts. Now all of them were there, abruptly thrown together in the same room. It had the discordant feeling of a dream where a familiar object appears in an unexpected setting, known but out of place.

And then I suddenly realized that something, or rather someone, *was* out of place. *Where's Timmy?* I asked the agent next to me. He scanned the room and then looked back at me, shrugging. *Come on, let's go,* I told him, and we dashed up the stairs to the second floor. We pushed open the first door we came to. Inside, Tim's head turned toward us. The teenager was sitting in front of a computer.

My first nightmarish reaction was that he had run up to the terminal to wipe the hard drive. But he made no sudden movement back to the computer, allaying my fears.

Are you Tim? I asked.

Yes, he nodded.

FBI. You need to come with us.

Tim quietly stood up, with myself and the other agent on either side of him. A computer forensic expert slipped past us, sat down in his chair, and quickly called the lab at Quantico to run down the computer's manufacturer and operating system and any quirks of the configuration to ensure we captured every bit of digital evidence. We walked Tim down the stairs to the living room, where his handcuffed parents were exiting separately into waiting cars to be taken back to the field office to be photographed and fingerprinted.

If Tim was surprised at a gaggle of FBI agents leading his parents away, he didn't show it. We took him to Alex, who was already with the social workers and the rest of the team who were going to help them figure out what to do with the next day, the next week, the next month. I stepped outside briefly into the Cambridge afternoon while the search team spread throughout the house, placing signs with numbers in each room and photographing them, documenting the state of things at the beginning of the search.

"DON'T CHOOSE BRITAIN AS A PLACE TO LIVE"

Many aspects of Ghost Stories remain classified, and I can't discuss our conclusions about what Tim and Alex knew about their parents' true identities. What I can say is that both children sued to retain their Canadian citizenship; while Canada conveys citizenship to individuals born on its soil, that right does not convey to the children of diplomats posted there. During the litigation, the Canadian government argued unsuccessfully that as the children of officers of the SVR, Tim and Alex should be considered children of foreign diplomatic personnel. In court arguments supporting that position, the Canadian government disclosed that it believed Tim knew about his parents' real identities and their covert work for Russia. Interestingly, the Canadian Supreme Court recently granted Alex Canadian citizenship, finding that he did *not* know about his parents' true identities prior to their arrest that afternoon in Cambridge.

Not surprisingly, neither Andrey nor Elena was willing to say much, and they were turned over to the U.S. marshals for detention. All the illegals were transferred to the Southern District of New York to face charges of being unregistered agents of the Russian government. After less than two weeks of high-level diplomatic negotiations, they boarded an unmarked Boeing 767 on July 9 and flew to Vienna, Austria. Four Russian prisoners waited on the tarmac there to board the plane for the return flight. It was a spy exchange — Russian spies for Russians who were alleged to have worked for the West. After the two groups swapped places, the 767 took off again, flying first to the United Kingdom to drop off two of the former prisoners before continuing on to Dulles Airport. After that, the passengers faded into anonymity in Western life.

That is, until March 2018, when the long arm of the GRU, the Russian military intelligence service, reached out from Moscow to execute an operation reminiscent of the most nefarious active measures of the Cold War. On a Sunday afternoon, passersby at an outdoor shopping mall in the English town of Salisbury noticed a balding older man and a younger woman slumped on a bench, slipping in and out of consciousness. They were alive but in a catatonic state.

The man's name was Sergei V. Skripal, and he was one of the Russian prisoners exchanged on the tarmac in Vienna in 2010 for the deported illegals. The young woman was his daughter, Yulia. Investigators concluded that the Skripals' mysterious illness resulted from exposure to a virulently powerful nerve agent known as Novichok. A chemical weapon Russia developed at the peak of the Cold War, it is believed to be one of the deadliest chemical weapons in the world, even more so than the extremely lethal V-agents, such as VX.

Skripal and his daughter both recovered, but the assassination attempt was widely viewed as retribution by Putin for the former Russian intelligence officer's having served as a double agent for Britain. The poisoning sent a chilling message to disloyal Russians: the SVR can find you, wherever you are, and will punish you and your loved ones. Putin's actions also again broadcast a message to government adversaries: Russia can still strike beyond its borders, exerting clandestine influence, carefully making moves in the great game that has outlived the Cold War.

As with the active measures of days past, Russia denied having anything to do with it. Russian foreign minister Sergey V. Lavrov called the accusations "nonsense." But as with all active measures, winks and sneers followed the official denials. Lavrov warned Russians not to betray their country, and if they did, "Don't choose Britain as a place to live." A newscaster on state-controlled Channel One warned that the "profession of a traitor is one of the most dangerous in the world." And as Putin coasted to reelection a week after the poisoning, he said that one misdeed could never be forgiven: "Betrayal."

I would face the same accusation from my own president for attempting to safeguard our nation against the very foreign enemy that had been trying to infiltrate and destabilize the United States for generations. Skripal was a traitor in Putin's eyes for betraying his country. I would become a traitor in Trump's eyes for defending mine.

Russia has been the alpha and the omega of U.S. counterintelligence throughout the 20th and 21st centuries. The Kremlin's active measures, the underhanded dirty tricks and *kompromat,* didn't die with the collapse of the USSR but instead continued clear through until today. But Mother Russia was hardly the only existential threat to our country. By the time we shut down the illegals ring in 2010, we were almost a decade into another case —one born of a tragedy that had consumed America's attention, to Russia's enduring benefit.

3

Sea Change

ON A LATE summer morning in 2001, I waited in my car by the curb at Logan Airport. A warm September breeze blew in through the car window, and jets thundered overhead in a crisp blue sky without a cloud in sight. Nine months after the cold January night of the Cambridge bank break-in, I still was in the thick of Ghost Stories. On that morning I was waiting for George, the meticulous agent who had helped us slip into Andrey and Elena's vault. George was arriving on a U.S. Airways shuttle with a man named Fred, one of his partners from the Flaps and Seals team. They were coming back into town to help troubleshoot a small issue that we had encountered while slowly drawing our net around the Russian couple.

Just before 8 a.m., George and Fred exited the terminal, heaving the familiar Pelican cases and backpacks like the ones they had used for the Harvard Square bank job. We shook hands, I popped the trunk for the bags, and we piled into the car. I merged into the rush-hour traffic looping around Logan and into the line of cars snaking toward the Sumner Tunnel to Boston.

Popping out of the tunnel into the sunlight of Boston's North End, I took Cambridge Street, which curved around the back of Beacon Hill toward the Longfellow Bridge and crossed over the Charles River to Cambridge. We were somewhere around Massachusetts General Hospital at about 8:45 a.m. when Fred's pager went off. Sitting next to me in the front passenger seat, he pulled his pager off his belt. It was, for that time, a fancy one: nationwide coverage and the ability to send and receive extended text messages. There were no smartphones yet, and BlackBerries wouldn't become ubiquitous for at least another year.

Fred reached into his bag and pulled out a pair of reading glasses. After he got them on his nose, he looked down at the tiny screen.

Weird, he said. *A plane just hit one of the World Trade Center buildings.*

Minutes later we were headed back to Logan, from which that plane — and another — had recently departed, hijackers aboard.

AND THE WALLS CAME DOWN

It was a simple question. *Will we get any remains?*

The husband and wife leaned against one another at the Hilton Hotel at Logan Airport, her shoulders bent in palpable grief as she spoke to us. The day felt like it had gone on forever, but it still wasn't even late afternoon. September 11, 2001, had turned into an incongruously beautiful day, with no need for air conditioning, so our interview room was quiet and still. The FBI had paired up with Massachusetts State Police officers, effectively doubling the number of investigative teams to track down leads, and we sat across from the married couple. We had gone through the introduction — *I know this is hard and I can't begin to imagine the pain of your loss, but we're trying to figure out who's responsible and bring them to justice.* And then, gently, the questions. *Why were they traveling? Were they alone? Did they call to say goodbye? Did they mention anything unusual?*

Some families, friends, and next of kin of passengers on the two Boeing 767s that had departed from Logan earlier that day had gone to the airport for a variety of reasons — to seek answers, to help, to grieve, or just because they didn't know what else to do. We were interviewing them all, for our own reasons, gathering information and evidence, piecing together the facts of what had happened. Universally, people wanted to help, and I saw how strong a positive mental benefit came from giving people a way to contribute in some way amid a staggering tragedy. Restaurants donated food, and a conference room at the FBI field office was converted to a kitchen, set up with rows and rows of Sterno burners to keep food warm. Florists gave flowers, clergy offered prayers.

September 11, 2001, irreparably changed our country. It wounded the nation's collective psyche, and it individually scarred anyone old enough to remember the sight of the Twin Towers collapsing, of the Pentagon's smashed west face, and of the furrowed earth in Shanksville, Pennsylvania, where Flight 77 disintegrated as it plowed into the earth. Thousands dead and missing. Nineteen perpetrators dead. The mastermind, Osama bin Laden, deep in Afghanistan, under the protection of the Taliban. A network of Internet-savvy radical Islamists spread across the globe, deftly manipulating technology to communicate and recruit new adherents. The crime scenes — the planes and the buildings — obliterated in a volcanic inferno. Evidence incinerated or crushed into the earth, pulverized into dust, and, in New York, belched out over the now hellish moonscape of lower Manhattan.

The September 11 attacks also dramatically changed how the FBI conducted counterintelligence and how the Bureau interacted with its sister intelligence agencies, including the CIA and the NSA. Crucially, September 11 also radically altered the digital landscape of counterintelligence and cyberwarfare through a cascading series of events that unspooled over the course of more than a decade. It would in time even affect how Russia conducted active measures against the Main Enemy.

But on that September morning we were still operating in an analog world. We were more concerned with finding scorched remains of passports and bits of luggage than with looking for digital fingerprints online or trying to decrypt messages sent through proxy servers. We were trying to untangle a crime that was inconceivably complicated, and we were only at the beginning, with what now seem like the bluntest of tools at our disposal.

Almost two years earlier I had had a grim preview of what was to come. On October 31, 1999, the first officer of EgyptAir Flight 990 en route from New York's John F. Kennedy Airport to Cairo locked the cockpit door, took control of the aircraft, and intentionally plunged it down into the Atlantic Ocean about 60 miles south of Nantucket. As the plane rocketed toward the ocean with the crew screaming from behind the locked cabin door, he murmured over and over, "I rely on God."

The exact reason for the suicide dive that murdered all 217 people on board has never been definitively proven. Working with the coast guard and the National Transportation Safety Board, the FBI investigated the event. I was still a probationary agent little more than a year out of Quantico, and so I worked late shifts or random work, the tasks probationary agents do. A massive hangar at Quonset Point, Rhode Island, began filling with debris, a giant jigsaw puzzle of the plane, where preternaturally knowledgeable Boeing experts examined the smallest scraps of metal, consulted blueprints, and sent the debris to the correct spot of the gridded hangar floor. A separate storage site held the personal belongings of the passengers, and a refrigerated facility served as the morgue. There is no gentle way to say this: things in a large aircraft — metal, luggage, remains — sometimes do odd and unexpected things in crashes. Not everything disintegrates; an intact seat and passenger might separate from an adjacent seat split into a thousand pieces. In part because I had a toddler at home, with another one on the way, the hardest things for me to see were the infant clothes and a baby bumper destined for a crib in a nursery's matching pastel colors, stripped by kinetic force from a suitcase that had been carrying the gift.

The EgyptAir tragedy, sprawling as it was, represented a crime scene in miniature compared to September 11. Yet I couldn't help but think of it as I sat with the grieving survivors in the interview room at Logan Airport.

The husband and wife looked at us, waiting for an answer. From my experience on the EgyptAir disaster, I knew the truth wouldn't be comforting: *I don't know.* The husband was surprised, almost angry — he seemed to feel there was no way anything would be recovered — but there was little else I could say. *I'm so sorry for your loss,* I said, inadequately, then repeated, *and I'm sorry, but I don't know.*

Everyone I know who responded on that day carries memories with them. That husband, wife, and their child are one of mine. I say a prayer for them every September 11, and write about them here so that they — and the more than 3,000 other people who lost their lives at the hands of al Qaeda's attack that day — know that they are not and never will be forgotten.

I spent the next several weeks at Logan, chasing down flight manifests, reviewing videotape of gate and check-in counters. Acting on a tip about suspicious activity in the parking garage that morning, my temporary partner — a marathoner and West Point grad — and I found a rental car that three of the hijackers had used to drive to the airport. When a local nightly news crew later appeared to get a shot of the car as we waited for a bomb technician to sweep it for explosives, I ducked behind a pillar to keep from being seen. Given my ongoing role in the illegals investigation and my general desire to remain anonymous, I had to walk away until the reporters got their shot and departed. Only then could we secure the car for processing by our evidence response team — who later found, among other belongings, notes that the attackers had left behind. (I won't describe them, in case they are ever used as evidence at military trials at Guantanamo Bay or criminal trials in the United States.) We tracked down the logical leads from the car, interviewing employees at the rental agency and reviewing hour after hour of video footage of the parking structure's entrances and exits.

My time in Boston provided me with opportunities to learn and watch leadership in action, as extraordinary men and women put emotion aside and methodically laid out an investigative plan to find out what happened. And then they did what the FBI does better than anyone else in the world: retracing the attackers' steps, uncovering their movements, communications, funding, spending, and coconspirators.

But while we worked to chase down leads and build a by-the-book investigation into the attacks, the nation was adrift in uncharted waters. A sense of outrage and dread blanketed the country. Part of that dread sprang from the widespread belief that the attack had come from nowhere, without any kind of warning or clue. Not true. As the public would later discover, the country also suffered from the understandable fear that terrorism of such enormity could happen anywhere. And it was an understandable fear that if it could happen on a clear September morning in New York and Washington, arguably the two most closely monitored and carefully secured cities in the country, then it could happen again. At a busy shopping mall in Minnesota. At a festival in Tennessee. At football games on college campuses or NASCAR racetracks or the Super Bowl or the NBA finals. Terror could strike anyone, at any time, anywhere.

The many postmortems of that day of horror, including the 9/11 Commission Report, concluded that legal and bureaucratic firewalls both within and between federal agencies had prevented information-sharing. Legal barriers erected in the 1970s after the Church and Pike Committee investigations of intelligence abuses had sharply curtailed domestic surveillance. The passage of the Foreign Intelligence Surveillance Act (FISA) in 1978, and the creation of the Foreign Intelligence Surveillance Court to oversee the granting of surveillance warrants, provided a check and balance of the judicial branch when we sought to conduct things like searches and wiretaps in national security cases. But the passage of FISA also began a division of some of the FBI's criminal and national security work into two different camps. Between them arose what came to be known as "the Wall," an unintended creation of procedural changes that attorney general Janet Reno had instituted in the 1990s. Originally related to structuring information-sharing, the regulations had the unintended consequence of impeding the flow of information between the criminal and national security sides of the FBI and the intelligence community. And then there were simply parochial conflicts, sniping, and professional jealousies among the various agencies responsible for national security.

Topping it off, there were basic failures of technology. The FBI had made do for years with outdated, clunky technology, with agents sharing computer terminals more suited for a pre-Internet age. When I arrived in Boston in 1998, two agents shared one computer. Fortunately, even then it wasn't that difficult to find a terminal, as many senior agents still relied on dictagraphs, recording their interviews onto cassettes that would be forwarded to the stenographers' pool to be transcribed.

The clear intelligence failures of September 11 caused an immediate reevaluation of these flawed practices and outdated laws governing the sharing of intelligence and interagency collaboration. The regulations had hindered our ability to put intelligence information in the hands of criminal investigators even in the face of indications that a catastrophic terror attack was imminent. They also failed to reflect the reality that the FBI's work — whether it concerned terrorism, counterintelligence, or organized crime — increasingly had a global scope.

Boston never felt the same to me after September 11, and neither did day-to-day work at the Bureau. The first days were surreal, with flights suspended across the U.S. and all roads to and from Logan Airport blocked to nonemergency traffic. I left for work well before sunrise and arrived home after dark. It would be weeks before I saw my children awake. And even when work returned to some semblance of normality, it was a "new normal" — a more frenetic, anxious, and fast-moving mode of existence than FBI agents like me had known in the pre–September 11 era.

Growing up abroad, I had sometimes sensed fear and uncertainty beneath a thin veneer of normalcy in the places my family lived. In the aftermath of September 11, I sensed a similar fear and anxiety beneath the picture of stable American life. It wasn't just the intelligence community that changed after September 11. So too did America — and with us the world.

On the day of the attacks, elected leaders hustled into undisclosed secure locations and asked what more was coming. The FBI and the U.S. intelligence community didn't have a good answer, and that uncertainty generated a demand to assess what we could and should do to change that.

The September 11 attack unleashed a blizzard of responses large and small, and no single book is long enough to chronicle them all. The invasions of Afghanistan and Iraq. The dispersal of intelligence officers and FBI agents around the globe to chase down every clue that would lead to the perpetrators and, hopefully, prevent future attacks. The cleanup at Ground Zero and the Pentagon. The operations, both overt and covert, to intercept conspirators. The attacks rightfully consumed the country, and in many ways still do today.

Congress, of course, was a part of this, ratcheting its legislative machinery into high gear. Acting with lightning speed, the House approved the

USA Patriot Act on October 24, 2001, the day after it was introduced. The Senate passed the bill the following day and sent it to the White House for President George W. Bush's signature. He signed it the next day.

The USA Patriot Act fundamentally enhanced the flow of information within the FBI as well as between the FBI and the rest of the government. A piece of intelligence developed by the CIA in Yemen could rapidly be put to use in an investigation in Chicago, while intelligence from a phone call monitored under FISA authority in New York could be quickly sent to soldiers in Afghanistan to take military action. The Patriot Act also dramatically changed the standards for issuing national security letters, or NSLs. Like subpoenas, NSLs require an entity, typically a business, to turn over certain types of information to law enforcement agencies.

NSLs had been in use since the mid-1980s, and when I started in the FBI, approval frequently involved months of sending typed communications back and forth between the field office and FBI headquarters. After the Patriot Act, the heads of individual FBI field offices, not just headquarters, could receive approval to issue them, and coupled with the automation of the NSL process, approval time was slashed to days or even hours. After the inevitable grumbling about the learning curve associated with the new process, we began to make routine use of NSLs, with much more agility and investigative benefit, in a way that had always been available to agents working criminal matters with U.S. attorneys' offices.

We desperately needed these new authorities, especially in the counterterrorism fight. In 2010, I watched without envy from my espionage squad at the Washington field office (or WFO, in Bureau shorthand) as my colleagues in the FBI Counterterrorism Division fought to identify and keep track of Islamic radicalization around the United States. Increasingly that was being done online. Anwar al-Awlaki, the Yemeni-American radical cleric, would have had little influence in analog eras past, but in the digital age his polished videos had a global reach that inspired lone wolves and al Qaeda recruits alike.

But as I watched, neither I nor any of my colleagues appreciated the broader lesson for our own work: the growing power of online media to influence people's beliefs and actions, and how that power might be used by nation-states rather than religious radicalizers. It would take another five years for the gravity of that threat to become clear.

The information age would revolutionize the way we thought about counterintelligence threats in another way, too. Rapid advances in data storage and sharing technologies were empowering a new breed of spy— and laying the foundation for some of the biggest intelligence failures the United States had ever seen. This challenge and the one from online media would burst into our consciousnesses at practically the same instant.

A WAREHOUSE ON A THUMB DRIVE

On June 7, 2013, a blockbuster article appeared in the *Washington Post*. The article purported to describe a top-secret National Security Agency surveillance program called PRISM. The article reported that the NSA and FBI were tapping the central servers of nine leading U.S. Internet companies to extract audio and video chats, photographs, email, documents, and connection logs, to help analysts track foreign terrorism and intelligence targets. According to the article, the program arose out of the warrantless wiretapping program that had caused so much controversy in 2007.

The *Washington Post* cited a "top-secret document" that the paper had obtained. Similar articles appeared two days earlier in the *Guardian* in Great Britain and eventually in the *New York Times* and *Der Spiegel* in Germany. It wasn't long before the source of the documents became public.

Just over two weeks before the *Post* article had appeared, on May 20, 2013, a slim, bespectacled NSA contractor named Edward Snowden had stepped off a jetliner in Hong Kong. Using encrypted communications, he had arranged a meeting with journalists Glenn Greenwald, Laura Poitras, and Ewen MacAskill to conduct an interview and turn over thousands of documents that he had downloaded while working for NSA contractor Booz Allen Hamilton. Using a thumb drive and deception, Snowden allegedly exploited his security clearance to steal classified material and transport it to another country, get it into the hands of the media, and have it broadcast to the world.

On June 21, two weeks after the *Post* published its first article based on Snowden's smuggled intelligence, the Department of Justice unsealed a two-count indictment against him for violating espionage laws and stealing government property. With his passport revoked and facing certain arrest in Hong Kong, he fled two days later to Russia, where he is still ensconced in a secret location as of this writing.

I watched Snowden's flight from the front office of the Counterintelligence Division and later supervised the investigators pursuing Snowden. Because his criminal case remains open, I cannot detail specifics of what we did or what the Bureau plans to do. What I can say is that the FBI has a long memory, the agents and analysts working on Snowden's investigation are talented and tenacious, and I believe that Snowden should return to the U.S. to face the charges levied against him in front of a jury of his peers in a system of justice and laws built over centuries by the American people. He will get a fair trial when he finally chooses to return.

As we went about opening our investigation into Snowden, we struggled to understand what, and how much, information he had allegedly stolen. But one thing was clear: he had been aided in his theft by recent advances in information technology.

In the famous saga of the Pentagon Papers, RAND analyst Daniel Ellsberg had smuggled thousands of pages of the Department of Defense's 47-volume study of the origins of the Vietnam War out of his office over many weeks. The copying of the top-secret history in turn required its own elaborate secret apparatus to get the documents to the *New York Times* and the *Washington Post*. Now a thumb drive stores gigabytes of information, solid-state drives the size of a pack of cards can hold terabytes of data, and server farms stashed abroad can receive vast quantities of information almost instantaneously — or be targets of hacker attacks.

We had reason to believe Snowden had taken an extraordinary amount of data. One question was what he had done with it, and as we began digging, we found worrying connections to organizations whose interests were counter to those of the U.S. Although professing to be advocates of transparency and neutral publishers of information, they were proving to be anything but. One of them was WikiLeaks.

In 2010, WikiLeaks was still a little-known Web destination that posted hacked or leaked documents. Its obscurity vanished when the site posted an explosive video in April 2010. The grainy black-and-white aerial footage, taken from the cameras on an Apache helicopter, showed a 2007 air attack that killed about a dozen people, including two Reuters employees. A group of men milling in a Baghdad street included a Reuters photographer carrying his camera, which one of the soldiers aboard the Apache believed was a rocket-propelled grenade, according to the audio that accompanied the footage. The graphic video showed the gruesome nature of a war in which combatants mixed with civilians with tragic results. WikiLeaks, which Australian journalist Julian Assange had founded four years earlier, rocketed to prominence.

A month and a half later the army arrested a 22-year-old intelligence analyst then named Bradley Manning, who had been stationed at Forward Operating Base Hammer in Iraq, about 40 miles from Baghdad. A tormented young soldier whose father had pressured him to join the army, Manning grew to oppose the war and leaked the video to WikiLeaks, which then posted it online to the world. But that was not all. Manning had also downloaded some 250,000 diplomatic cables and more than 475,000 intelligence reports on the wars in Iraq and Afghanistan, which WikiLeaks published throughout 2010 after sharing some of them with the *New York Times,* the *Guardian,* and *Der Spiegel.* Manning, who identified as and eventually transitioned to female and took the name Chelsea, was tried by the army and received a sentence of 35 years in prison in August 2013. President Obama commuted the bulk of the sentence in 2017.

The Manning episode thrust WikiLeaks and its founder, Julian Assange, onto the global stage. It also threw them into Russia's orbit — and did so at a critical moment. While America's national security apparatus was focused on al Qaeda, Russia had been taking full advantage of its Main Enemy's preoccupation with another adversary. Moscow was moving to strengthen its position in the long competition with the United States and beginning to prepare its own hostilities against us. WikiLeaks, and many fringe media organizations like it, fit perfectly into these plans.

In 2010, just before Manning gave the video to WikiLeaks, the Russian government had launched RT America. Russia Today, founded five years earlier, had become a broadcast news powerhouse around the globe, winning awards and gathering kudos for its probing, inquisitive reporting, which sometimes verged on conspiracy theorizing. The network often positioned itself as a provocateur, adopting the advertising slogan "Question More," as it used the cloak of objective, evenhanded journalism to peddle controversy and foment political division. RT found a willing audience globally among those opposing the war in Iraq and broader U.S. foreign policy goals.

Debate is healthy, necessary, and firmly a part of the democratic process, but there were curious aspects of RT that suggested it was something less than an objective journalistic outlet. For instance, the network always had a decidedly anti-American bent. In 2012 the network began carrying a show by Julian Assange, in which the WikiLeaks founder interviewed prominent scholars, politicians, activists, and hackers, including Hassan Nasrallah, the leader of the Iranian proxy group Hezbollah, whom Assange praised for fighting "against the hegemony of the United States." The following year the network launched a provocative ad campaign called "Second Opinion." Prominently displayed in the Washington, D.C., and New York City mass transit systems, the ads showed artistic renderings of British prime minister Tony Blair, U.S. secretary of state Colin Powell, and President Bush with the banner "This Is What Happens When There is No Second Opinion," followed by "Iraq War: No WMDs 141,802 Civilian Deaths." The campaign intimated that the media in the U.S. and Great Britain were monolithic and shut out meaningful debate on issues like the Iraq war. RT, the ads claimed, presented an alternative.

The fact that RT was acting at the direction of the Russian government, serving as a clandestine propaganda arm, was suspected but not known at the time. Faced with the threat of prosecution for illegally acting on behalf of Russia, RT ultimately registered as a Russian agent in November 2017, a full year after the 2016 presidential election had been decided.

While it may have been a new phenomenon to some, the use of propaganda to influence opinion is nothing new. When people are aware of it, it tends to be because of clumsy attempts to portray an obviously biased point of view. But when propaganda is done effectively, people don't know that's what it is, and it can be a means to sow discord, to divide and destabilize our country and others. And in Russia, the United States was encountering an adversary with decades of experience in propaganda and disinformation — and that now, like Snowden and Assange, had the tools to match its ambitions.

CHESTS IN THE LAKE

Boston gave me my first bank break-in. It also provided my first taste of Russian disinformation.

During the second Russia-Chechen war, a local university professor named Lawrence Martin-Bittman reached out to our office. Martin-Bittman was conducting research for an article about Russian disinformation and wanted to relay some interesting events he had observed involving the manipulation of media coverage of the Chechen conflict.

To all appearances, Martin-Bittman was a journalism professor at Boston University whose work frequently focused on international media. He had a particular zeal for Soviet and Russian use of disinformation. Like Don and Ann, however, Lawrence Martin-Bittman was not his real name.

Born Ladislav Bittman in Prague, Czechoslovakia, in the early 1930s, Bittman had joined the Czech intelligence service. There he had received training from the KGB, becoming a specialist in active measures. The term is a KGB catchall expression to describe influence activities ranging from disinformation to acts of violence up to assassinations, all intended to sway political and social views or opinions. Social engineering, in other words, geared toward creating perceptions that aid Russian national interests.

Bittman was, in effect, a dirty trickster for the Soviets. In 1964 he became the Czech government's deputy commander of the Department for Active Measures and Disinformation. One of his most elaborate and infamous schemes, which was intended to cast aspersions on West Germany, involved sinking German military chests filled with papers in a Czech lake, to be "discovered" by a television documentary crew filming at the lake. An experienced diver, Bittman personally led the crew that retrieved the underwater chests. After they were opened, the Czech government reported that they contained Third Reich records that showed that former Nazis had spied for West Germany after World War II. All planted, all forged, all intended to discredit West Germany.

After the brutal suppression of the Prague Spring uprising of 1968, Bittman grew deeply disillusioned with the Soviet Union. He defected to West Germany, then received asylum in the United States. With a new identity as Lawrence Martin, he eventually settled in the Boston area. In 1974 a Czech military court convicted him of treason in absentia and sentenced him to death. Twenty years later, following the Soviet Union's fall and the Warsaw Pact's collapse, his sentence was lifted, after which he made his true past public, appending Bittman to his surname. Scores of aspiring journalists passed through his doors to learn how to detect and use disinformation.

I called Martin-Bittman, arranged an appointment for later that day, and hopped on the MBTA, using the subway to avoid the headache of trying to park in the clogged streets around Boston University. I walked into his office alone and spent some time chatting amiably with him. I knew he was sizing me up, assessing me, but it was only later that I appreciated how talented he was. At holiday parties with our CIA colleagues, we were used to a punch list of standard questions intended to not-so-subtly size us up. *So, what do you do? How long have you been in the office? Where'd you go to school?* It's a practice we referred to as "case-officering," as in *Look, asshole, stop case-officering me. Let's get another glass of whatever cheap wine you're pouring and talk about the Patriots.*

But Martin-Bittman was doing nothing of the sort. As he chatted with me, he drove the conversation without my feeling that he had taken control of it, in the graceful way that truly athletic runners don't appear to expend any effort. He was highly competent and completely unremarkable, a version of what is known as a "gray man" in the intelligence world. Someone who can work a source without being noticed. Someone who can vanish into a crowd. Someone who is there, expertly doing his job, but doing it so inconspicuously that an untrained observer doesn't remark on it, afterward having only the vaguest recollection that someone was there at all. If the target is asked to remember details, there are none; the person is simply a gray man in the background of the memory.

When Martin-Bittman called up the FBI to talk, Russia was deep into its military offensive against Chechnya, justifying its actions in part by pointing to atrocities and terrorism that the Russian government attributed to Chechens. Martin-Bittman had been following the Russian media coverage and was struck by what he saw as the same active-measures techniques and modus operandi he had been taught by the KGB decades before. I don't recall precisely how he presented his observations to me — perhaps he had digital recordings on his office computer, or a TV with a VCR that he alternately played and paused — but he walked me through Russian television coverage of Chechnya. The news reports showing civilian casualties in the bloody conflict seemed glossy and smooth, but Martin-Bittman claimed that his trained eye could see inconsistencies, which he pointed out to me. Slight changes in video resolution and lighting at critical junctures suggested that the images of injured civilians had been recorded at different times or in different locations, he said, then spliced into the Chechnya war footage. Then Martin-Bittman stepped back to explain to me how the material supported the strategic Kremlin narrative. It had been carefully placed into media reporting in a way, he argued, to hide the true hand of Russian intelligence behind the footage.

I don't remember the meeting being particularly revelatory. The U.S. intelligence community already knew the Russians were making use of propaganda in the Chechen conflict. Although I was fascinated by the meticulous care with which the disinformation had been slipped into the stream of public knowledge, it seemed a bit superfluous and ham-handed, like the work of scissor-wielding Soviet propagandists who had outlived their era. I wasn't sure of its value, and I was far from impressed. The Soviets, then the Russians, placed great value on active measures, but to me it didn't seem that the return was worth the investment. While disinformation might work in oppressed societies, Western openness — including our aggressive and independent press — seemed a strong inoculation against propaganda.

Fifteen years later I would come to understand how deeply I had misjudged the power of disinformation, and I would kick myself for not having seen what Martin-Bittman foretold and what it might mean in a digital era. Long before the 2016 presidential election, the Russians were deeply and openly involved in social media manipulation in a variety of internal and regional elections. As a community, we did a poor job of understanding the implications.

For generations of counterintelligence agents, disinformation meant fabricated news reports or altered photographs. It meant bogus academic research and whisper campaigns. It meant propaganda; it meant chests planted at the bottom of a lake. Mostly it meant getting disinformation into the hands of the press, academicians, or politicians, who unknowingly disseminated the information to the public.

Today there has been a sea change in how Russia runs active measures. A skeptical press, gullible politicians, and easily impressed academics are no longer required to distribute false information. The Internet has created an entirely new cloak for clandestine work. Thanks to Facebook, Twitter, and any number of other social platforms, disinformation, leaked email, and fake news can reach millions of people instantaneously, distributed anonymously by amplifying networks using proxy servers, virtual private networks, and IP-masking technology to hide their tracks. A Facebook friend request from Professor Smith in Norman, Oklahoma, might actually be from an academic at the University of Oklahoma, who just happens to really like intriguing but ultimately inaccurate articles from Sputnik — another international news agency connected to the Russian government — about chemical attacks in Syria. Professor Smith might also be a GRU subcontractor with the Internet Research Agency in St. Petersburg.

Not only that, the information can be spread again and again and again, kept alive on the Internet, bouncing from news feed to news feed, chatroom to chatroom, tweet to retweet. Moreover, in the social media environment, disinformation doesn't need a strong and sophisticated push to inject itself into the dialogue: millions of users already inclined to seek out and read proof of what they believe lend their power to spreading the bogus information.

Active measures feed on uncertainty and doubt — feelings that were all too common in the United States in the period after September 11. Our fears intersected neatly with Russia's strategic interests. America's collective psyche, Moscow was discovering, was quite a target-rich environment.

In 2012 an overseas tragedy linked to the September 11 terror attack took place in Libya amid a maelstrom of chaos and violence that would, over time, pump endless fuel into conspiracy theories, bitter acrimony, and domestic division within the United States. The Arab Spring had begun the previous year with immense optimism as protesters began to seek democratic change across the Middle East. But hope that change would arrive peacefully soon collapsed. While a few countries saw peaceful democratic reforms, others slid into violence and bloodshed. Libya was one of them. As civil war engulfed the country, Muammar al-Gaddafi was deposed and killed. Then, on the anniversary of September 11, a mob gathered outside the U.S. consulate in the Libyan city of Benghazi, stormed the poorly secured compound, and killed four Americans, including ambassador J. Christopher Stevens.

I can't and won't attempt to relitigate the complexities of that event, but I bring it up for one reason: in 2014, the Republican-dominated U.S. House of Representatives voted 232 to 186 to create the Select Committee on Benghazi to investigate what had happened and determine if there was any culpability on the part of secretary of state Hillary Clinton, whom Barack Obama had defeated in the presidential primary of 2008 and who was widely assumed to be the 2016 presidential candidate for the Democrats.

Bitter debate over Benghazi raged through the end of 2012 and 2013, consuming our leaders' attention while elsewhere dark stars were beginning to align. Far from the toxic political environment in Washington, Donald Trump held the 2013 Miss Universe Pageant in Moscow. While he was there, he met with a Russian real estate developer about building a Trump Tower Moscow. That same year, a Romanian hacker who used the name Guccifer quietly hacked and released email from accounts of American celebrities, members of the Bush family, former secretary of state Colin Powell, and Sidney Blumenthal, an informal adviser to Hillary Clinton who had swapped dozens of messages with Clinton about Libya. Guccifer, who was eventually arrested and extradited to the United States, later claimed to have hacked the email of Clinton herself (although he later confessed to us that he had lied about having successfully done so).

A few weeks after the House voted to launch the Benghazi investigation, four Russian men and women applied for U.S. visas to travel to the United States. They were friends, they said, who had met at a party and wanted to travel together. Of the four, two women, Anna Bogacheva and Aleksandra Krylova, received visas. Their plane touched down on U.S. soil on June 4, 2014.

They were not, as they claimed, friends who had met at a party. They were employees of the Internet Research Agency, a St. Petersburg–based organization that received funding from Russian oligarch Yevgeny Prigozhin, an ally of Putin's. The job of the shadowy IRA "troll farm" where they worked was the equivalent of sinking fake Nazi papers in a lake, updated for the 21st century. To sow chaos, discord, and uncertainty across the Internet, dozens of IRA "specialists" operated fake accounts and personas on various U.S. social media platforms, focusing on Facebook, YouTube, and Twitter, eventually adding Tumblr and Instagram accounts. One of the IRA's early "achievements" was concocting a fake chemical disaster in Louisiana. Dismissed at the time as a sick prank, the hoax was a sophisticated cyber operation spread across numerous social network platforms, using fake websites, manufactured screen grabs, and text messages.

With the nation's political establishment fixed on the Benghazi scandal and its divides, Bogacheva and Krylova set out across the United States. They had a whirlwind itinerary, a road trip that would take them to nine states in three weeks. They took snapshots along the way and stopped to chat with friendly Americans about the things that were on their minds. Perhaps they talked about the direction the country was headed. Perhaps they talked about the Black Lives Matters movement. Perhaps they talked about Benghazi and the terrible tragedy that had taken place there. Perhaps they talked about the presidential election just two years away. Perhaps they talked about the woman who was likely to be the Democratic forerunner.

At the end of the trip they boarded planes back to Russia, taking with them all the intelligence they had gathered. And then they got to work.

PART II

4

Midyear Exam

WASHINGTON SLOWS IN August. The summer heat and humidity drape the city like a wet blanket, the streets empty out as residents flee for vacation, and the pace of the city sputters to a crawl. But not for the FBI.

On the morning of Friday, August 21, 2015, I drove from my home to the Washington field office eager for my day to begin. I had been assigned to WFO since 2014 as an assistant special agent in charge, the second in command of an FBI's field division, running all of the office's espionage cases. I was also overseeing proliferation cases, investigating anything related to banned weapons of mass destruction or sanctions-busting conventional weapons.

WFO has a unique place in the U.S. intelligence community. The FBI office that conducts investigations in the Washington, D.C., area, WFO is distinct from FBI headquarters, which oversees all the FBI's various operations. Its local counterpart in DOJ is the U.S. attorney's office in Washington, which under law prosecutes any espionage committed overseas by U.S. citizens. That prosecutorial jurisdiction conveys itself to WFO's investigative reach. As a result, WFO agents run some of the most interesting and complicated counterespionage and counterintelligence cases around the globe. In 2015, WFO's recent caseload included NSA leaker Edward Snowden, Iranian proliferation networks, and Jerry Lee, a former CIA officer who would plead guilty in 2019 to conspiring to commit espionage for China (a case that I had authorized opening years before in my earlier position as a WFO squad supervisor). The office's jurisdiction for international casework attracted the very best agents, analysts, and supervisors, making it easy to build an exceptional team for every case. The work was exciting, the people were excellent, and their talents were superb. I was content. And all of it was about to be upended.

After I reached my desk, I got a call to bring a supervisor to headquarters for a briefing on a sensitive case. The briefing topic wasn't mentioned. But as I hung up, I had a pretty good idea that it was connected to rumors I had been hearing around WFO about a new investigation involving Secretary of State Clinton.

In the summer of 2015, the Benghazi Select Committee was delivering constant fodder for headlines and nightly news broadcasts as House Republicans performed a slow-motion flaying of the State Department for the violence at the consulate in 2012 that had killed four Americans: Ambassador Stevens, Tyrone Woods, Glen Doherty, and Sean Smith. They had quickly focused on investigating Secretary Clinton, who was already the Democratic frontrunner for the 2016 presidential election.

Members of the media and the public also began demanding information about what had happened. Congress and Freedom of Information Act requests forced the State Department to collect and release Secretary Clinton's email. And as it did, the department's inspector general noticed, early in the summer of 2015, that some of the email messages contained what appeared to be classified information.

That discovery alone might not have been terribly newsworthy. So-called spills of classified information — the inadvertent disclosure of classified information in an unclassified email — happen routinely in government. What made this different was the fact that Clinton had been using a private email account to conduct business, not the State Department's official system.

Department of State regulations, both written and verbal, were vague and contradictory when it came to employees' email protocol. It's not clear that Clinton *had* to use the State system. What made it a case for us was classified information appearing on that private account. The State Department shared this information with the intelligence community's inspector general, who in turn referred the allegations to the FBI, on July 6. Four days later the FBI had opened an investigation out of headquarters.

I wasn't involved in the case at that point, but I had heard rumblings about something related to Clinton's email and the FBI examination of them. The focus of the investigation wasn't clear to me, and moreover, it wasn't obvious that it was a major effort. If it had been, I would have heard more about it. It would have been next to impossible to conceal details from agents and analysts trained to uncover hidden work, who have the same appetite for interoffice intrigue found in every workplace.

I called Wayne, one of my squad supervisors, to go with me downtown to headquarters, about a half-mile away. Wayne was a no-nonsense agent, a barrel-chested former college wrestler from Ohio who had spent the bulk of his career working on drug and violent criminal matters and leading WFO's SWAT team. Though new to counterintelligence, he ran a squad that included agents and analysts with deep experience in cases involving U.S. Army leaker Chelsea Manning, WikiLeaks founder Julian Assange, and U.S. Marine Corps general James Cartwright, who later pled guilty to lying to the FBI about disclosures of classified information to the media.

Judiciary Square, where WFO is located, is only a few blocks from headquarters, but the heat was oppressive, so we hopped into my car for the short drive. We parked in the cool subbasement near spots reserved for bucars — FBI shorthand for "Bureau cars" — and took the elevator from the 10th Street NW and Pennsylvania Avenue NW entrance up to the fourth floor, which housed the Counterintelligence Division.

When we walked into the main conference room, I was pleased and surprised to see an old friend named John, an assistant United States attorney in the Eastern District of Virginia whom I had worked with for a decade. Pleased because he was extremely skilled with classified criminal work. Surprised because his presence meant FBI and DOJ headquarters had already decided to include a local U.S. attorney's office. In counterintelligence work, most cases did not reach that threshold. His presence told me that the investigation had reached a stage that required some sort of criminal process, like a subpoena, a search warrant, or grand jury testimony, or was close to it. The fact that he had traveled to FBI HQ also told me that the investigation would be using those tools within days.

Soon after Wayne and I arrived, the head of the division, Randy Coleman, walked in with members of his leadership team and shook hands with everyone in the room. Like Wayne, Randy was an old-school agent who spoke with a straightforward manner that inspired respect. Then I got my second surprise: the appearance of a senior analyst named Derek.

I had worked with Derek frequently for more than a decade. A diehard New York Yankees fan, he was tall and acerbic. Derek had a unique blend of intellect and practical operational experience. He knew his subjects cold and was considered a top analyst within the Bureau and across the intelligence community. Though neither of us knew it then, we would first partner on the Clinton investigation and then continue working together for months after that on an even bigger case involving the secretary of state's soon-to-be rival for the presidency.

We settled into our seats as the briefing started. *Thanks for coming over, folks,* Coleman began. *Before we begin, I need to stress to you the absolute need for secrecy about everything we're about to talk about. It has to remain locked down, no exceptions.*

His next words dispelled any lingering confusion about why I was there. Derek and I had been reassigned to take over the investigation into whether a top presidential candidate had mishandled classified government email. The candidate, of course, was Hillary Clinton. The case was called Midyear Exam.

THE BUBBLE

In the weeks leading up to this moment, I had been focused on the cases in front of me, but it was impossible to ignore what was happening elsewhere in the country. When I had driven to work earlier that same Friday morning, hundreds of people had already lined up at a football stadium in Mobile, Alabama, hours before an evening speech by an unconventional presidential candidate: a reality TV star whose campaign had been a side-show until it somehow seized center stage.

Donald J. Trump had waded into the presidential race on June 16, 2015. Descending on an escalator in Trump Tower in Manhattan, he waved and flashed double thumbs-up to the crowd. When he reached the flag-draped dais where he would make the speech announcing his candidacy for the highest office in the land, he made it immediately clear that he had a far different message from Obama's mantras of "Hope" and "Change We Can Believe In."

Trump painted a bleak picture of America in decline and under assault, "a dumping ground for everyone's problems." He singled out Mexicans coming over the southern border: "They're bringing drugs, they're bringing crime, they're rapists." Beset by foreign adversaries and dangerous immigrants, he said, the United States was a nation without victories. Shooting wars, trade wars, any kind of war — America was losing.

Trump, of course, asserted that he was the remedy. He touted his reputation as a gold-plated dealmaker who could turn the tables on trade adversaries. The United States, he said, needed to extricate itself from foreign entanglements and stop playing the role of global sucker, footing the bills for allies overseas. At campaign events, the crowds roared as he tossed insults, belittled opponents, and scoffed at political correctness. "America First" was his creed. "Make America Great Again" was his slogan.

I didn't see his announcement speech, but I read about it that night. I instinctively recoiled, as many Americans did. I was incredulous that he would make such outrageous claims. Shocked that he would target an entire group of immigrants as criminals. Troubled by the dog whistles to racists and conspiracy theorists. And as he later gained his party's nomination, I was amazed at his refusal to release his tax returns or commit to relinquishing his business interests if elected. Voters expect candidates and politicians to rise above personal biases and self-interest. Trump didn't seem to have the desire or need to do any such thing.

At the time, it just felt unseemly. Few people, myself included, saw the momentous upheaval it portended. I didn't yet have an inkling that Trump's casting aside of presidential norms would draw me into an investigation of this unpredictable future president, or of his likely Democratic opponent, Hillary Clinton. I would never have imagined that stadiums packed with Trump's fervent supporters would ring with thundering chants of "Lock her up!" aimed at Clinton and her handling of email. The very email that I was now in charge of investigating.

By the time I was summoned for the August briefing, Midyear Exam was up and running right in the heart of D.C., at FBI headquarters. The Counterintelligence Division had brought in several agents on a temporary basis to start looking into the allegations of mishandling that had emerged from the Benghazi inquiry, pairing them with some analysts permanently assigned to headquarters. As with any sudden influx of personnel, the new arrivals were sprinkled wherever desks could be found, their names plugged into the phone system and their seat assignments sent to the mail-room.

Agents have a name for cases run out of the Hoover Building: Head-quarters Specials. Though WFO received its share of high-profile espionage cases, on rare occasions the FBI preferred to have the most sensitive coun-terintelligence cases run out of the downtown HQ. Doing so was intended to provide immediate senior-level direction and control over the cases, as well as to insulate those investigations from the rest of the Bureau.

The men and women of the FBI, like the employees of the CIA and other agencies in the intelligence community, are really good at keeping secrets. But they are human — and people talk. Leaks, whether they touch on the lives of sources buried deep in Moscow or Beijing or on the investigation of the presumptive Democratic nominee for president, can lead to disaster. In time, indeed, the threat of leaks would prove to be as challenging as any penetration effort the Russians or Chinese could have mounted.

Running the Midyear Exam investigation out of the Hoover Building must have seemed like a wise precaution, and an adequate one. But while the principle behind running high-priority cases out of headquarters seemed sound, experience had taught me that it didn't always work well in practice.

Years before, when I was a squad supervisor at the Washington field office, I had inherited what had been a complex Headquarters Special, an investigation into disclosures of classified information to then–*New York Times* reporter James Risen. The case included sensitive details about the early stages of the U.S. warrantless surveillance program, which was called Stellar Wind at the time. When I received the case file, years after the investigation had begun, it was a mess. Interview reports hadn't been incorporated into it. Unlogged evidence had been stored in safes scattered around FBI headquarters and WFO. File reviews, a process by which supervisors gauge the progress of an investigation, hadn't been conducted. Challenges like these were endemic to Headquarters Specials, and I knew I would need to find a solution.

The difficulty was that headquarters simply isn't suited to conducting investigations. Unlike FBI field offices, which are manned by agents and analysts (who investigate cases), surveillance personnel (who follow subjects), and technical experts (who provide tools like those to monitor calls, collect video, and record conversations), staff at FBI HQ lacks the day-to-day expertise and structure essential to good casework. Compounding that problem, no one outside the Hoover Building is looking over the shoulders of agents working on Headquarters Specials. Field offices have their own chains of command separate from headquarters, with approvals required at every investigative step that the field agents take. That independent accountability functions as a kind of check and balance, which is absent from cases out of headquarters — a distinction that would become especially meaningful in cases that were sure to be highly politicized, like the Clinton investigation.

With Midyear Exam, I didn't have the option of moving the case to Washington Field, so I did the next best thing: I brought Wayne and most of his WFO squad over to headquarters, everyone from agents to analysts to support personnel. While we would work out of the Hoover Building, I wanted to replicate the operational structure of a field office investigation, augmented with extra analysts and computer forensic experts. Another advantage of this arrangement was that Wayne and his squad knew the DOJ players very well, the prosecutors from both the U.S. Attorney's Office for the Eastern District of Virginia and the National Security Division at DOJ who would be our daily partners for most of the next year.

The FBI and DOJ have a complex, codependent, and often fraught relationship that is challenging to understand from within, let alone describe to someone on the outside. DOJ is the parent agency of the FBI, and the FBI director reports to the attorney general. But DOJ is helmed by several layers of political appointees, whereas the FBI has only one — the director — and is mostly led by career intelligence and law enforcement people with no discernible partisan bias. Prosecutors within DOJ decide which FBI cases are prosecuted and which aren't. Despite this subordinate structure, the FBI fiercely guards its political and operational independence. It's almost never an antagonistic relationship, but it is always a complicated one.

Space at headquarters is as jealously guarded as schematics for an atomic bomb. Early on, we had no luck in finding four cubicles together, let alone the 15 spots that we would need for the core investigative team. As I searched for secure space large enough to house our team, an informal connection — a friend at the Strategic Information and Operations Center, or SIOC — helped get things done, as is often true with bureaucracies.

Deep in the heart of the Hoover Building, SIOC is closest to what the public envisions a high-tech FBI command post to be. In the multiroom complex, large-screen monitors with video feeds display real-time operational activities and cable news broadcasts. Secure video conferencing equipment hooks into a dedicated White House system. Rooms filled with rows of computer terminals serve to coordinate responses when multiple national incidents unfold simultaneously.

In the center of it all, an octagonal room with a raised center platform and glass walls has a 360-degree view of the command post. Called "the Bubble," it could have been lifted straight out of the Fox thriller *24*. The room had served as the original nerve center for the command post, but by 2015 SIOC staff had outgrown it and moved to a larger room.

Thanks to my friend's intervention, the Bubble was ours. And truly, we couldn't have asked for a better space from which to be running this highly sensitive investigation. The Bubble was private and secure, behind multiple series of doors requiring badge access. Most headquarters employees didn't have access to SIOC; the Bubble served as an isolated enclave away from prying eyes. We covered the lower parts of the windows for even more security, and in early fall began moving into the space that would be the team's investigative home for the next ten months.

Both Derek and I wanted a fast pace for the investigation while ensuring that team members shared information quickly and efficiently about the multiple investigative threads. To make sure that that happened, we held mandatory briefings every morning in the Bubble to walk through every aspect of intelligence and operational updates and forensic and legal issues. We wanted a formal information-sharing structure, but we also wanted some peer pressure to reduce or eliminate reports of "no update today," which is uncomfortable to deliver when hardworking colleagues next to you are detailing steady progress and results. Truthfully, we didn't need to do that — without exception, aggressive overachievers made up our team. It quickly became obvious that this was a self-motivating squad. If there were to be a problem, it would be managing confident, ambitious all-stars who rightfully felt they were the best people for any particular job.

It was clear to me from the beginning that the overall investigative strategy required three distinct avenues. The first involved understanding how Clinton used email and locating as much of it as we could find on devices or servers. The second was to figure out who had placed classified information into the email messages and why they had done it. And the third was to determine whether any unauthorized person — from a foreign nation to a hacker — had gained access to the classified information.

While easy to describe, the first task — mapping out, then tracking down Hillary Clinton's email — proved extremely complicated to execute. To start, there wasn't just one server that housed Clinton's email, contrary to the popular impression of the case that would emerge later. Before I arrived on the team, Clinton's attorneys had negotiated with the Department of Justice to turn over the famed computer that had served as the private server operating her private email account from her residence in Chappaqua, New York. We dubbed that the "Pagliano server," for the staffer who had set up the system. What Clinton's attorneys neglected to mention was that her email domain had been transferred to a professional hosting company named Platte River Networks, or PRN, based in Denver, Colorado. I was suspicious after our forensic analysts discovered in September that Clinton's email was being hosted on a different computer from the Pagliano server. The attorneys hadn't lied to us, because no one had asked, "Are there any other servers?" We didn't know we needed to. But our goal was pretty clear — we were trying to find all the places where her email had been stored or sent while she was secretary of state. Given our obvious investigative goals, why hadn't her attorneys told us?

Whatever the reason, we quickly obtained access to the PRN server. That revealed new complications. The physical server wasn't maintained at PRN's headquarters in Denver but instead at a server farm in Secaucus, New Jersey, across the Hudson River from Manhattan. On a rainy Friday evening, a group of agents and computer forensic analysts drove up to New Jersey, suppressing grumbles about a soon-to-be-lost weekend, and checked into a hotel near the gray environs of the Meadowlands. The next morning we interviewed officials at the server farm, a company called Equinix. When we made our way into the hardware cage, we found not only PRN's equipment but additional hardware used on the original Pagliano server. And there was more. Unbeknown even to Clinton's lawyers, the new server also used a storage device that was making remote cloud backups to a company in Connecticut.

One discovery kept leading to another. And beyond all the places where Clinton's email was stored or maintained — the servers, backups, and so on — we also needed to find every device she had used to access it. The Black-Berrys. The iPads. The laptops. Any equipment she used in her office at the State Department and her residences in Washington, D.C., and Chappaqua. All the hardware. Luckily for us, it turned out that the secretary of state was not very tech-savvy. Clinton didn't use a computer at the State Department at all, only her ubiquitous BlackBerry. She didn't even have a computer at her State Department desk.

After we had figured out how Clinton had been using email and located all of her messages on these various servers and devices, this first part of our investigative strategy wouldn't be complete until we finished another time-consuming task: tracking down recipients of all her email to get copies. Every one of Clinton's email messages had been sent by or to her. We quietly put out a request across the federal government to search for and collect any email Clinton had sent or received. We had to make sure that we weren't missing anything, on the off-chance that for some reason not all of Clinton's email was still stored on her own servers and devices.

Once we found all her email, our second investigatory thread was to determine what information in it was classified. Like locating the email, this was a huge undertaking. Because Clinton had used her account for both official and personal purposes, a variety of privileged communications, such as correspondence with her attorneys, former president Clinton, or medical professionals, had to be removed by a separate FBI filter team walled off from the investigative team. After they reviewed every email we found, they sent the work email to the investigative team. Next, another set of analysts reviewed each of those, identifying the ones that might contain classified information. Then we sent those to the various U.S. government agencies who might have a say in the matter, to ask them if any of the contents of the email was in fact classified.

Sharing the Clinton email with the various stakeholders in the U.S. government was a monumental effort in and of itself, because in many cases multiple agencies needed to weigh in on a single email. A hypothetical example: a U.S. ambassador overseas sends a foreign minister's reaction to a U.S. military action that includes some U.S. intelligence assessment of the impact on the foreign country. That email would be flagged for review by the Departments of State and Defense as well as the CIA. We did that for thousands upon thousands of email messages, most of which did not end up being classified. But we needed the agencies who "owned" the information to tell us that; we couldn't determine it ourselves.

After we knew which email messages were actually classified, the second investigative track required us to determine why classified information had ended up in what was supposed to have been an unclassified email, and the intent of those who had included it. In most instances the classified information had been sent to Clinton — most often forwarded by third parties — rather than written by her. What was the purpose of the email? Had the sender tried to write it in a way to make it unclassified? Was there an exigent circumstance, a decision in which someone's life was at risk? Determining any of those was a herculean task that required tracking down people around the world, from the State Department command center to U.S. embassies and consulates abroad, and sending agents to interview each person.

The third and final investigative stage was to determine if any of the classified information had reached unauthorized parties, whether from carelessness, hacking, or some other means. If that had occurred, we needed to determine what to do about it. A lot of the work we did as part of this third track remains classified, but it involved a host of experts at the FBI Laboratory in Quantico, Virginia, and around Washington, D.C.

It wasn't until months into the investigation that we began to get a sense of the information that would allow us and DOJ to adequately assess whether we could prove each element of a crime in a way that would support bringing charges. But by the approach of winter, as agents fanned out across Washington and boarded planes across the U.S. and around the globe to interview individuals who had written the classified components of the email in question, and as we began picking apart the stories behind each message, two things quickly became apparent. The first was there had clearly been mishandling of classified information. The second was that we weren't finding evidence of behavior that the Department of Justice had traditionally charged for violations of criminal law.

At one point I remarked to Derek that I doubted that we would even have opened a case if Clinton's activity had taken place on a State Department server instead of a private one. He agreed. The experienced agents and analysts on the team saw it too — the various members of the team had collectively worked scores of mishandling investigations. One of them quietly remarked, *It's a shame we can't have a team like this on a bona fide espionage case.* It wasn't that Midyear didn't deserve our investigative efforts; rather, it was a sense of disappointment that we didn't have the resources to commit to major counterintelligence cases under investigation at the same time that involved Russia, China, and other countries. Those cases at least involved bona fide threats from a foreign adversary. Midyear Exam was proving to be what would otherwise have been more of an administrative issue.

Various House committees focused with laserlike intensity on the question of whether classified information had been mishandled. The assumption was that this question had a simple yes-or-no answer. But it's actually not as cut-and-dried as it might seem. Tens of thousands of people have clearances in the U.S. government, and classified information is inadvertently put into unclassified email on a regular basis, just as Trump's UN ambassador Nikki Haley did years later. Most of these spills, although violations of rules governing the handling of classified information, are treated as security incidents and are dealt with in an administrative manner, not a criminal one.

Only a tiny number of mishandling incidents meet the threshold for the FBI to open a case. Each month my espionage section would receive mishandling allegations from other U.S. government agencies, only for us to reject them because of a lack of any indication of criminal intent. Indeed, the intent of the person involved was a significant consideration, one that hinged on several factors. Was an unauthorized party, such as a journalist or a foreign national, the recipient? How much material was involved, and how sensitive was it — was it megabytes or boxes of highly classified information, or something less? Finally, how did the person behave after mishandling the information — did he or she lie to us or destroy evidence? Or did that person behave normally, apart from the violation itself?

If the FBI has a high bar for opening mishandling cases, DOJ has an even higher bar for prosecuting them. While the FBI is an independent investigative agency and guards that independence fiercely, it is still a part of DOJ, which decides whether or not to bring criminal charges in a case. The FBI can and does advocate to prosecute a case, but DOJ makes the call.

I remember numerous unsuccessful arguments to get DOJ interested in cases of hoarders who took home hundreds of pages of secret documents, or angry ex-spouses who called to tell us about boxes of classified information

in their basements. Based on that history, I had a strong hunch that DOJ
would not find the evidence strong enough to bring charges in Midyear.
With every case, DOJ weighs a variety of factors in deciding to prosecute.
Part of that puzzle is a statutory gap in the law as it applies to mishandling of
classified information. The law didn't apply neatly to previous mishandling
cases, and it didn't line up exactly with Clinton's behavior either. While U.S.
law has a powerful statute for regulating the knowing disclosure of national
defense information in the form of the Espionage Act, DOJ has almost always used it in cases of espionage involving a foreign power: for people like
Aldrich Ames, Robert Hanssen, and Jerry Lee. Historically, DOJ has been
loath to diminish the law's power with minor cases, especially those where
ill intent was lacking, which would carry a number of unwanted effects,
such as establishing a courtroom history of light sentences. Most mishandling cases are treated as misdemeanors, if they're even considered criminal
cases at all. As an example, the FBI doesn't investigate, and DOJ doesn't
prosecute, most violations of the Federal Records Act.

Everyone on our team had worked on cases of mishandled classified information and had experienced their frustrations. And as we labored through
Midyear, our findings and past experiences made it apparent that we'd have
a hard time reaching the threshold for criminal charges in this case.

The biggest obstacle was our inability to prove intent at a level consistent
with previous cases in which DOJ had brought criminal charges. In each
case in which classified information appeared in Clinton's email, it was in
the context of people doing their job — it wasn't being sent to someone who
shouldn't have had it, like a reporter or a foreign intelligence officer.

Even the most sensitive information found in the Clinton email, from a
compartmented program classified at the top secret level, would have been
difficult to build a prosecution around because it appeared in the context of
legitimate work. I can't say more about it, because the information remains
classified to this day, but suffice it to say that all the prosecutors and agents
with whom I worked this case also saw the same problems with trying to
base a criminal charge using that top secret information.

That spill of information from the top secret program was the outlier;

the overwhelming majority of the mishandled information in Midyear Exam was confidential, the lowest level of the classification guidelines. By that, I mean it wasn't the type of information that would lead to the compromise of a highly placed human source or the identification of some very sensitive satellite technology or communications interception capability. Historically, trying to get DOJ to prosecute mishandling of confidential-level information was like trying to persuade a fraternity house that spiked seltzer is a real drink. And in many cases within Midyear, the material had been classified *after* the email had been sent, a process we referred to as "up-classifying." A diplomat overseas, for example, sends a summary of a public meeting she had with a foreign government official. Someone in D.C. looks at it and decides that it might hurt the U.S.'s national security if it were released: a foreign official might be embarrassed and unwilling to talk to us again, or be offended and unwilling to work with us. As a result, the email would get classified after the fact in order to protect the information. Here's another example: email sent by former government officials or other private individuals can also be deemed classified after the fact, even though the people writing the email do not have security clearances, or don't have them any longer. In still other instances, some information that was marked as classified was found upon review not to be classified at all.

In short, it was a mess. But not, apparently, an illegal mess.

We weren't finding evidence that Clinton or anyone on her team had deliberately destroyed work email to avoid turning it over, an action that otherwise might have led us to recommend that DOJ pursue criminal charges. We were getting additional copies of the email we'd received from the Clinton camp from all around the government — from people who emailed with her and all the places those email messages had been forwarded. Sometimes we received three or more copies of the same email chain, or portions of it. Those copies substantiated the email we received from Clinton's team. We just weren't uncovering a large number of work email to or from Clinton that she had not already turned over to our investigators — there were some, but not many — and this aboveboard behavior was another strike against a showing of criminal intent.

We did discover one email deleter, and he was a PRN employee. Long before the Benghazi Committee, Clinton's team had asked him to delete nonwork email from the server. He forgot. It was only after Congress had issued a subpoena for Clinton's email that he realized in a self-described "oh shit" moment that he had never deleted them. So he went in on his own and did so. When we came to talk to him, he initially denied it — a bad decision that brought DOJ's prosecutorial discretion into play. Had he lied to us? Yes, as he later admitted under a proffer agreement (which limited our ability to use his admission against him) provided by DOJ. Was he part of a grand conspiracy with the Clintons to obstruct the investigation? Certainly not. He denied receiving any instructions to delete email beyond the initial request, and when we confronted Clinton's team with what happened, they were visibly surprised. They didn't know what he had done until we told them. Scared and suffering from some questionable legal advice, he eventually came clean. He had possibly broken the law, but would justice have been served in a scenario in which the only person who ended up being prosecuted was a hapless IT guy who got caught up in a hyperpartisan environment while Clinton walked away without being charged? The prosecutors didn't think so, and neither did I or the rest of the team.

Eventually I would recognize an aggravating irony in this investigation. Among the many reasons Clinton took so much heat for using private servers and personal email as secretary was the real danger that a foreign nation could hack into her server and access her email. We didn't find any digital fingerprints that suggested foreign hacking or snooping, though nations like Russia and China are good enough to break in without leaving a trace, especially if we get to the evidence well after the fact, when logs and other files have been overwritten or deleted. But the idea that Clinton's use of a personal server put sensitive information at risk nevertheless was a big part of the recriminations heaped on her by the Benghazi Select Committee and others.

The fact is that if Clinton's email had been housed on a State Department system, it would have been less secure and probably much more vulnerable to hacking. During 2015 hackers repeatedly breached U.S. government computer systems. Not just the State Department, which seemed to be constantly ejecting unwanted intruders, but the Pentagon and the White House as well. Derek and I joked about how her private email probably was more secure than a State Department system, which we *knew* would have been hacked.

While we investigated whether the secretary's email had been compromised, the man who would emerge as her GOP opponent appeared to be in the process of actually compromising himself. In September, while Trump was on the campaign trail, he gave his attorney and hatchetman Michael Cohen permission to quietly pursue the creation of Trump Tower Moscow, a huge Russian real estate deal potentially worth hundreds of millions of dollars in which Trump would license his name and brand. We didn't know this at the time, of course. Nor did we know that we would soon move on to a case in which a foreign entity had hacked not just email but our entire democracy.

Two months later, in mid-November, a new Twitter account appeared called @Ten_GOP, purporting to be the "Unofficial Twitter account of Tennessee Republicans." Trump's son Donald Trump Jr. followed it, and eventually he, campaign manager Kellyanne Conway, and campaign digital director Brad Parscale all retweeted various messages from the account. In reality the account proved to have nothing to do with Tennessee Republicans other than its bogus name. Created by Russian hackers connected to the Internet Research Agency, the account posted everything from manufactured information about crowd sizes at Trump campaign rallies to conspiracy theories about President Obama. Trump wanted to go to Russia, but the Russians had already come to America.

THE END OF THE BEGINNING

As revelers in Washington rang in 2016, most of the Midyear Exam team had reached a conclusion: we were unlikely to establish sufficient proof to bring criminal charges against Hillary Clinton for mishandling classified information. I was privately stewing over another issue, which was the need to devote such significant resources — in both quality and quantity — to Midyear. The agents and analysts working on this investigation all knew the other work that was out there: bona fide allegations of U.S. government employees spying for Russia, China, Cuba, Iran, and others. Because of Midyear, these investigations had fewer resources than they should have had. We all knew we had to investigate Midyear thoroughly and completely, like every other open investigation; nevertheless, it sometimes felt like we were unavoidably stuck on a politicized endeavor when we could have been spending our time and effort on cases that would actually make America safer.

The vast resources dedicated to Midyear weren't for the purpose of investigating the espionage case of the century; this clearly wasn't that. Rather, the time, money, and manpower poured into our investigation were aimed at more prosaic, political goals: to conclusively document what Clinton had or had not done while demonstrating the FBI's credibility. In late December, Trump had retweeted an image of a woman wearing a "Hillary for Prison 2016" T-shirt; Clinton's email was clearly on the minds of many voters, pundits, and politicians. I think all of us knew that we were building an airtight case to withstand the political storms to come. But I don't think anyone, including myself, had any idea how devastating those storms would be.

Over the winter of 2015, the investigation had advanced to the point at which we were ready to interview the principal players. We would need to talk with Clinton's top advisers, such as Jake Sullivan, Cheryl Mills, and Huma Abedin. Eventually we would need to talk to Secretary Clinton herself. I felt confident that we were on a path to wrap up the investigation by early spring. We were certainly aware that the Democratic primaries began in February, but that wasn't a milestone that we were pacing ourselves with. Instead, the political race going on in the background was like a relentless drumbeat, growing ever louder as we moved the case to a conclusion.

And then everything came to a halt. As we reviewed our forensic analysis and talked with Clinton's attorneys, we discovered the existence of two laptops that no one had previously mentioned.

We were furious. We didn't yet know it, but the revelation would stretch the investigation several months longer than we had anticipated, putting us on a collision course with the upcoming Democratic nominating convention and thrusting our investigation even further into the spotlight than we could have anticipated.

The reason the laptops were important was this: the production of Clinton's email was conducted by two attorneys on two laptops that were in the possession of a third attorney. Eventually we called them the "sort laptops," because they were the original computers used by Clinton's team to segregate State Department email messages from the personal ones.

Independent of our discovery of the laptops, a sharp-eyed analyst on our team made an observation that threw the sorting process into question. The analyst noticed that many of the work email messages we found from outside sources had blank or ambiguous subject lines. This led the analyst to what proved to be an accurate hunch: Clinton's attorneys hadn't reviewed every email, only elements of them, like the date, sender, recipient, and subject line. In other words, their sorting process had been flawed, and a small chunk of work-related email had not been turned over by Clinton's team. This wasn't nefarious on its face — it wasn't evidence of ill intent or proof that the new email contained classified information — but it forced us to reexamine the process and pursue other email that might still be out there.

We needed to get the laptops. We knew that at one time they had held all of Clinton's email, including work material we now knew had not been produced. We also knew that it was likely we already had copies of most, if not all, of the relevant email from our government-wide searches, but nevertheless this was an obvious and necessary lead for us to pursue.

I naively anticipated that DOJ would obtain the laptops fairly quickly. During earlier stages of the investigation, the back-and-forth between the Midyear team and Clinton's lawyers over consent for searching computer equipment had been slow and occasionally aggravating, but it had yielded results.

But this time proved different. The two Clinton attorneys who conducted the sorting, Cheryl Mills and Heather Samuelson, were not represented by Clinton's firm, Williams & Connolly. Instead they hired a well-regarded and combative former prosecutor who had been part of the Oklahoma City bombing prosecution team. We should have known this spelled trouble based on our previous experience with her; she had represented Jake Sullivan, a Clinton foreign policy adviser whom we had asked to interview earlier in the investigation — a request that had been met with stonewalling. Eventually the prospect of a DOJ threat of a grand jury subpoena had broken the logjam, and Sullivan had appeared for a voluntary interview. But our new encounter with this attorney would not end so happily: before long, negotiations between the DOJ and Clinton's attorneys over the laptops collapsed. Claiming a variety of legal privileges and ethics obligations, Mills and Samuelson's attorney refused to allow them to be interviewed about the sorting and would not consent to turning over the laptops.

I was irritated by the delay. Even though we believed the information on the laptops was unlikely to change our understanding of the case, reviewing them was obviously a necessary investigative action. To credibly conclude an investigation, you simply must take certain steps, or at least attempt to take them. Analyzing the laptops was one of those steps. We knew that, DOJ knew that, and Clinton's team, with all its various legal factions, knew it. In my mind, there were three possible reasons they were digging in their heels: either the left hand of the Clinton team wasn't talking to the right hand, or they were taking what I thought was a very shortsighted stand about the principle of privilege, or they were concerned enough about something on those laptops that they were willing to go to the mat about it. Whatever the reason, I didn't like it.

Winter of 2016 turned to spring as the Midyear investigators wrangled with our own attorneys and the Midyear attorneys argued with Clinton's various attorneys. Despite the legal warfare, we began winding up other aspects of the investigation that were not contingent on the laptops. We did those interviews that didn't hinge on what may or may not have been on the sort laptops, interviewing Huma Abedin on April 5, 2016, in a conference room at the Washington field office. Then we settled into a holding pattern that would last for almost three months.

And as we waited, the threat of a foreign hack — the very specter that had drawn so much attention to Clinton's email in the first place — became a reality. The beneficiary of the breach would be the one person who had benefited from the Clinton scandal more than anyone else: Donald J. Trump.

5

Pandora's In-Box

THE DAY AFTER we interviewed Huma Abedin, on April 6, 2016, a staffer at the Democratic Congressional Campaign Committee opened up an email and let in the Russians.

The Benghazi Committee, it turned out, wasn't the only group interested in Clinton's email. Starting in March 2016, official-looking notices had arrived in the in-boxes of hundreds of Democratic National Committee (DNC) and Clinton campaign staff members. The messages appeared to be security notifications from Google, urging the staff to reset their passwords using a link embedded in the text of the notice. The email messages were actually a spearphishing attack with spoofed Gmail addresses. They came from a specialized military cadre called Unit 26165 within Russia's Main Intelligence Directorate of the General Staff — the GRU — and included a hyperlink that led to a GRU website that harvested the email passwords of the duped account holders. An earlier email had gone to Clinton campaign cochair John Podesta on March 19. He had promptly clicked on the link. Two days later the Russians siphoned roughly 50,000 email messages from Podesta's account.

After the DNC's IT department detected the intrusion, the DNC reported the hack to the FBI. I learned about it well after the fact, in the late summer of 2016, as we scrambled to put together the pieces of what was unveiling itself as an enormous Russian attack on our election.

At almost the same time that the Russians first tunneled into the DNC system, they registered a Web domain as DCLeaks.com, using a service that anonymized registrants' identity. DCLeaks would go live on June 8, one day after Clinton declared victory in the Democratic primaries in California, Montana, New Jersey, New Mexico, and South Dakota. The following week the site would post the stolen DNC email as well as hacked records from George Soros's Open Society Foundations. In a kind of nod to bipartisanship, the website would include a small number of stolen GOP email messages. The "About" section of the site said that DCLeaks "was launched by the American hacktivists who respect and appreciate freedom of speech, human rights and government of the people." There was no hint that the site was part of a Russian active measure.

The release of email over the Internet — a bombshell development that would wrench our attention away from the Midyear investigation — was intended for as wide an audience as possible, with the goal of swaying the U.S. presidential election away from Hillary Clinton, who was the presumptive Democratic nominee. At the same time, alarming signs appeared that the Russians were working through other channels to reach one person in particular: Trump.

In early March 2016, the Trump campaign appointed George Papadopoulos as a foreign policy adviser. A 28-year-old energy consultant, he had unsuccessfully offered his services to the Trump campaign, then worked briefly as an adviser to then-candidate Ben Carson. In the winter of 2015 he took a job at the London Centre of International Law Practice and renewed his bid to work for Trump.

This time his efforts paid off, and the campaign — which was desperate for foreign policy experts — named him as an adviser. Told by Sam Clovis, the Trump campaign's national cochair and chief policy adviser, that improved Russia-U.S. relations were a top policy priority for the Trump campaign, Papadopoulos began to work on arranging a meeting between Trump and Putin. Had it come off, such a meeting would have been unorthodox, but not an invitation to illegal activity — unlike what actually happened next.

Papadopoulos had no Russia connections, but someone else at the London Centre did: Joseph Mifsud, a Maltese academic and the center's director. Mifsud had been dismissive of Papadopoulos when they first met in Italy. But when he learned of Papadopoulos's new campaign role, Mifsud became acutely interested in the ambitious young adviser.

Years later Papadopoulos would make a variety of inconsistent claims that what happened next was a setup, including misguided allegations that Mifsud actually was an asset of the Italian intelligence services, which used the Maltese professor — at the behest of the CIA — to compromise Papadopoulos in a "deep-state" attempt to undermine the Trump campaign. Trump's defenders have parroted this conspiracy theory. But it is just that: a debunked theory that appears to have been advanced to serve Papadopoulos's and Trump's political agenda — but also Putin's.

In any case, no debate exists about what happened next: Mifsud arranged for Papadopoulos to meet a Russian woman — whom he falsely claimed to be Putin's niece — with connections to top Russian officials. In late March, Papadopoulos flew to Washington for a campaign meeting, at which he described his new Russia connections and how he could facilitate the desired meeting between Trump and the Russians.

After the meeting, his efforts seemed to gained steam. On April 11, the Russian woman emailed Papadopoulos that "we are all very excited by the possibility of a good relationship with Mr. Trump. The Russian Federation would love to welcome him once his candidature would be officially announced." Mifsud also introduced Papadopoulos to Ivan Timofeev, another Russian, who claimed to be connected to the Russian Ministry of Foreign Affairs, the equivalent of the State Department.

On April 26, Papadopoulos met Mifsud for breakfast at a London hotel. Mifsud said he had just returned the previous day from Moscow, where he had met with high-level Russian officials. And Mifsud said something else: he had learned that the Russians had dirt on Clinton — "thousands of emails," as Papadopoulos would later tell us in an interview.

The next day Papadopoulos emailed the Trump campaign. "Have some interesting messages coming in from Moscow about a trip when the time is right," he wrote to Stephen Miller, a campaign adviser who would stay on with the Trump presidency and become an architect of the administration's immigration policy.

Papadopoulos's omission of details about "dirt" in that email was odd. Our best guess was that he deliberately avoided mention of it in email, or he withheld mention of it as leverage to increase the campaign's interest in him. That same day he emailed Corey Lewandowski, another top Trump campaign official, "to discuss Russia's interest in hosting Mr. Trump. Have been receiving a lot of calls over the last month about Putin wanting to host him and the team."

A few days later Papadopoulos emailed Mifsud to thank him for his help in setting up the potential meeting between the Russian government and the campaign. "It's history making if it happens," he wrote. His assessment was correct, but for altogether the wrong reasons.

THE SPEECH

If the FBI had not been so occupied with the Clinton email investigation, would we have spotted the Russians' active measures in early 2016? The answer is probably no, but the question nevertheless haunts me to this day. It's painful for me to reflect on the fact that while the Russians were busily working against us, we were investigating one of our fellow Americans in a case that revealed poor behavior but found no prosecutable wrongdoing.

Throughout the spring, Derek and I relayed developments about Midyear to FBI director James Comey, who closely followed our progress. While Comey would never say so, I sensed that pressure on him to conclude Midyear Exam was mounting — even if it only came from his political sixth sense. On May 5, Comey's chief of staff called me to convey the boss's desire to wrap up quickly. During a meeting later that day with Andrew McCabe, the deputy director, I jotted a frustrated note: "We're burning daylight." Time was running out and we needed to finish.

Neither Derek nor I relayed those sentiments to the team. From Comey down, all of us understood that we would take as long as we needed. At the same time, everyone also understood that the primaries were wrapping up and Clinton was looking more and more likely to secure the Democratic nomination. Among our team there was a universal desire to steer clear of the election. We were walking a tightrope, and I had to be careful to keep the priorities balanced. So I tried hard to strike the right tone: Do it quickly, but do it well. Hurry, but don't rush. The professionalism of the people on my team made that seemingly contradictory guidance easy.

There were signs of continued progress. For instance, on May 16, Comey asked if we were ready to interview Clinton, something we had been waiting to do until the seemingly interminable debate about the sort laptops resolved itself. We were. I called DOJ to begin planning the process.

Yet as summer 2016 approached, we still didn't have the sort laptops. To this day, I don't understand the delay. The Clinton camp still refused to provide them, but their attorneys were too good not to understand that the FBI wasn't going away. And then there were the optics: even if Clinton won a protracted legal battle, the spectacle of her refusing to surrender the laptops would likely be an untenable campaign position, given the almost daily public controversy over her email. We were concerned that DOJ might not have conveyed the seriousness of the matter to Mills, so we had told her ourselves during a prickly interview on April 9. While that blunt message caused some friction between the FBI and DOJ teams, it did force the issue to a resolution, with DOJ committing itself to obtaining the various internal authorities needed to subpoena the laptops.

I can only assume that the stonewalling was a simple miscalculation of the Clinton team. When our DOJ attorneys asked for assistance in getting the laptops, Clinton's lawyer David Kendall just shrugged. *There's only so much I can do,* he told them. I didn't buy it. I have little doubt that if Clinton had told Mills to turn over the laptops, it would have happened overnight. It was in the middle of the presidential campaign, and perhaps her counsel decided it was below the threshold of something she needed to weigh in on. Or perhaps it was a strategy: Clinton's team let Mills's and Samuelson's attorney be combative and create delay while Clinton's own attorneys maintained goodwill. Or perhaps they thought we would let the matter go, or delay it until after the election. Instead it just made us more adamant about seeing what was on the laptops.

When DOJ finally obtained approval to issue subpoenas in late May, the remaining pieces quickly fell into place, and we interviewed Samuelson and reinterviewed Mills at the end of the month, specifically about how they had conducted the sorting process. We also dug into the sort laptops. After all that fight, an analysis of the laptops revealed nothing new. As we expected, neither the interviews nor the reviews of the email altered our sense that that there were no grounds for criminal charges. This confirmation of our suspicions made the delay all the more baffling. Clinton's team had refused to submit to a basic investigatory step until threatened with subpoenas. For nothing.

Ultimately, the apparent obstructionism just fed Clinton's opponents' narrative that she had something to hide and that the investigation was dragging on because the FBI was somehow complicit. I had a sinking sense that the needlessly drawn-out investigation would only add fuel to the conspiracy theorists' fire, if and when we determined to recommend to DOJ that it decline to prosecute Clinton. This concern, as it turned out, would later prove justified.

Between what he was hearing from us and DOJ, and based on his own long experience as both a prosecutor and a senior DOJ official wise in the ways of Washington, Comey also anticipated that the case might end with no charges. He too seemed to have felt a sense of foreboding about how this would be perceived by the American public. Although the Clinton interview would not happen until early July, Comey clearly wanted to prepare for the eventuality that it would not fundamentally change our assessment that we should not recommend that DOJ charge her with a crime. In fact, before we had even scheduled an interview with her, Comey was considering how to explain such a potential decision to the American people — especially an electorate riven by partisanship and riveted by the ongoing FBI investigation of a leading presidential candidate. Always prepared, he began working on a speech that would deliver the news that Clinton would not face charges, and why.

On May 6, I opened an email from McCabe with a draft of the speech. McCabe was looking for feedback on behalf of Comey. I knew Comey was a gifted communicator. Still, I was stunned as I read the draft at my desk. My first impression was how complete it was. During briefings, Comey took few — if any — notes, leaving his black felt-tip pen and pad of paper untouched on a small stand to his right at his conference table. And yet he had drafted from memory an almost perfect summary of the case, his analysis of the facts, and his conclusions.

Much has been written about Comey, but it's worth adding a few words here. To someone outside the FBI, it might seem odd that Comey would seek input from the comparatively lower-level members of Midyear's leadership about a public statement and its content, especially given the regimented culture within the FBI. In fact it was not unusual. At the top of a deeply hierarchical bureaucracy, he understood the danger of homogenous group-think. The FBI's seventh floor, where the senior ranks cluster, was already physically isolated behind locked glass doors. That in turn created the risk of intellectual segregation, becoming an echo chamber of stale consensus. Comey tried to avoid that by encouraging a diversity of opinion and dissent. He liked debate, provided it was intelligent and respectful, and said he welcomed challenges to him and his senior leaders. I believed him. He would sometimes throw inflammatory ideas into a briefing, then watch and listen as the debate unfolded. During discussions of particularly complex issues, he would become quiet, staring down at his lap and absently fiddling with the tip of his tie, immersed in thought.

In short, while the draft of Comey's speech was already impressive, we knew that the boss wanted it to be even better. As the investigation dragged on, slowed by the maneuverings of Clinton's lawyers, we went over every word of Comey's speech, dissecting every idea, checking every punctuation mark. There wasn't much that needed correcting, and while we pressure-tested the underlying argument of the speech, we found that it aligned very closely with our understanding of the investigation and its probable outcome.

One principal conclusion that Comey wanted to impart was that, at the time he drafted his remarks, the facts of Clinton's case appeared to fall far short of the threshold set in previous DOJ prosecutions for mishandling classified information. In the theatrics of the campaign, the accusation that she had gotten away with a crime was a common refrain among Clinton's critics. *Folks, if you or I did what she did, we'd be in jail.* From what we could tell at the time, that was simply untrue, and if our final determination was that Clinton had not committed a prosecutable crime, Comey wanted the public to know it. Someone who mishandled information in the way Clinton appeared to have done — as a cabinet official with round-the-clock job requirements and national security stakes at play — would almost certainly face political pressure to resign or even be fired. But in the history of the statute, as far as we knew, no prosecution for actions like hers had ever been undertaken. Comey wanted to make that clear — assuming that a rigorous analysis of the case law did indeed support that conclusion.

Midyear prosecutors and investigators were comfortable with the conclusion that Clinton's actions did not merit charges, but with someone of her political stature, even that well-reasoned assessment wasn't enough. At the FBI we investigated alleged violations of the law, but we didn't prosecute those who broke it. So when Comey asked us to provide a list of every case in which the espionage statute had been applied to mishandled classified information to confirm our belief that Clinton's actions did not merit charges, we had to admit that we did not have that information. To get that list, he had to turn to DOJ.

The prosecutors were not happy. They grumbled because it entailed a lot of work and could provide a map of past practice for us in the FBI to use later when arguing for prosecution in other cases, but they agreed.

The data that DOJ eventually produced supported our recollections of prior investigations: DOJ had never charged mishandling cases similar to Clinton's. In fact, DOJ only brought charges in four broad categories, which Comey laid out toward the end of the speech — instances involving clearly intentional and willful mishandling, extremely large quantities of information, behavior indicating disloyalty to the U.S., or active obstruction of justice by doing things like lying or destroying evidence. Clinton's case didn't fit any of those categories, a revelation that lent strong support to our conclusion that she should not be charged with a crime. A speech from Comey was becoming more and more likely.

While parsing his speech, the team made a wording recommendation about Clinton's intent. In the first draft of his speech, Comey used the phrase "gross negligence" to describe her behavior. Shortly after he circulated the speech for comment, several of us were sitting in my office assembling various operational and legal comments. One of the attorneys noted that the phrase "gross negligence" carried a specific legal meaning; moreover, it appeared in one of the subsections of the Espionage Act. When I asked for a nonattorney answer to what "gross negligence" meant, I got a long, head-spinning response. One thing was clear: there was no settled answer. Case law defined the phrase in varying and sometimes conflicting ways. Furthermore, there was significant concern that the passage of the espionage law that used the phrase might be unconstitutionally vague, because it avoided requiring a demonstration of the person's intent when conducting the act. For that specific reason, DOJ had almost never applied that part of the statute in a prosecution.

The problem was that Comey wasn't writing the speech for lawyers or a law journal; rather, his remarks were intended for the American public. If he tried to explain precisely what he did and didn't mean by using the term "gross negligence" — or any other complicated terms with legal weight — he would plunge his audience down a juridical rabbit hole that his audience might never be able to climb out of. So we tried to figure out a nonlegal way to convey the actual gravity of Clinton's behavior fairly without invoking an ambiguous legal term that probably didn't apply.

After lengthy discussion, one of the attorneys suggested an alternative, which we all settled on: "extremely careless." Was that the perfect phrase? Maybe not. It was accurate and fair, but it also gave rise to a host of conspiracy theories that we had changed the language of the speech to "let Clinton off the hook." That's nonsense, and while I have heard Comey speculate that in retrospect he ought to have used a different term (and while I understand that feeling), I don't have a better substitute. Any attempt to characterize improper behavior succinctly and completely is invariably imperfect. At least the phrase we chose steered the debate around a legal minefield that wasn't warranted by the facts.

These discussions about the wording of Comey's speech also pale in comparison to a bigger, thornier, and much more consequential debate that also took place: about whether Comey should publicly comment on the Clinton investigation at all. As a team, we were not as focused on this question in May and June as we were later, in the fall, when we found ourselves grappling with the question of whether to notify Congress of the eventual reopening of the Clinton case. At the time our focus was on the outcome of the investigation and how best to serve the American public's interest given the notoriety of the case and the seemingly conflicted position of senior officials at DOJ who would normally announce the result. That's not to say that Comey didn't seek our input about the speech, or that we didn't debate the pros and cons in earnest. We gave him our honest opinions. It's simply that in retrospect, I wish we had spent more time challenging ourselves and our debate-loving boss to think even harder about his decision to provide an extended explanation of why we were declining to charge Secretary Clinton.

In the past, DOJ and the FBI had publicly discussed high-profile cases to explain the decision-making around criminal charges. For example, Comey noted in a later editorial that DOJ had released information about the investigation into the killing of Michael Brown in Ferguson and, before that, the decision to hold American citizen José Padilla in military custody as an enemy combatant after his arrest on suspicions of aiding terrorists.

But those cases were rare, and moreover, the FBI had never publicly commented without DOJ approval on such a significant case that wasn't going to be prosecuted. And on that point things got sticky. The public needed to have faith in the FBI's objectivity and the independence of its findings. But from the beginning, the case had taken on political overtones that added to partisan furor. Attorney general Loretta Lynch, the head of the Department of Justice, had asked Comey to refer to the case as a "matter" rather than an investigation, and President Obama had commented, mid-investigation, that he didn't "think [Clinton's email] posed a national security problem." Both episodes inflamed criticism that the investigation was being subjected to partisan influence. Comey's concern — a valid one to me — was that having DOJ involved in any public comment about this would only fan the flames.

Comey's concerns about the perception of partisanship grew ever more acute as the November 8 general election edged closer. On June 7, Secretary Clinton secured the Democratic Party's nomination for president, formalizing a victory that many observers had come to view as inevitable and officially pitting her against Donald Trump, who had won the GOP nomination the previous month. Now that the subject of the FBI investigation was officially one of the two presidential finalists, our efforts came under attack from both sides of the aisle. Trump seized on the investigation as a campaign rallying cry, stating during a speech on June 2, "Folks, honestly, she's guilty as hell. She's guilty as hell . . . It's a disgrace to our nation. It's a disgrace. So we'll see. We'll see what happens. I don't know. I've always had great confidence in the FBI. I must tell you. I have great respect. I know some FBI folks. I've always had great confidence in them. I can't believe that they would let this go."

And then something happened that further deflated our hope that the investigation and its findings would be seen as impartial while also shoring up Comey's sense that he needed to make a speech. On June 27, a plane carrying Attorney General Lynch parked on the tarmac in Phoenix, Arizona. Bill Clinton was aboard a private plane about 20 to 30 yards away. When Clinton learned that Lynch was parked close by, he disembarked, met her on the tarmac, and they stepped into her plane to talk casually until one of Lynch's staff frantically intervened. But the damage was done: the fact that Clinton had met privately with the attorney general while his wife was under investigation by the DOJ created the unshakable impression that the couple was attempting to influence Lynch, and at the very least an appearance that she was receptive to their overtures. For partisan critics of our investigation, this was further proof that any case not ending with a prosecution was hopelessly biased and could not be seen as a thorough investigation. Trump seized on the event and quickly drew in the FBI, tweeting on July 2 that "it is impossible for the FBI not to recommend criminal charges against Hillary Clinton. What she did was wrong! What Bill did was stupid!"

From my perspective, Comey had already committed to a speech, provided we uncovered no new groundbreaking information in the final days. We had been crafting it for weeks at that point, and the tarmac meeting served to confirm rather than precipitate the decision. Lynch's ill-advised decision to meet with Bill Clinton solidified the sense within the senior leadership of the FBI that a public statement from Comey without the involvement or approval of DOJ was the right path.

Finally, there was a still-classified matter that posed a challenge to the appearance of DOJ's objectivity. Earlier in the year we had received intelligence regarding allegations about political forces at play behind the scenes at the White House and DOJ. While some media reporting has speculated about its content, the government has never described it, and I'm not going to do so here. The bottom line is that while we didn't assess the underlying content of the intelligence to be credible, that didn't matter — it is fair to say that it would have created concerns about the motives behind DOJ's decision not to prosecute.

Having said that, and with the impossibly unfair benefit of hindsight, I now believe it was the wrong decision for Comey to give that speech. Of course, none of us could possibly have predicted that we would be forced to reopen the investigation, which occurred later that year, on the eve of the presidential election. We couldn't know at the time that Comey's speech would set a precedent which before long would come back to haunt us — and help Clinton's opponent.

THE INTERVIEW

Comey's final decision about whether to charge the secretary of state with a crime, and the speech that would convey that decision, still hinged on

Clinton's interview. If in the course of her conversation with us she revealed something that changed our understanding of her case, Comey might need to change the contents of his speech — or not give it at all. We didn't expect that to happen, but we had to be sure.

Does that mean that the outcome of the investigation was a foregone conclusion? Absolutely not: the accusation that Comey or any of us had made up our minds not to charge Clinton before her interview — that, in Trump's words, "the system is totally rigged and corrupt" — is simply untrue. But any good prosecutor or investigator in the course of a long investigation will develop a sense of where the case is going. Either you have the evidence to prove beyond a reasonable doubt that a crime took place or you don't. Our team knew we didn't, and almost as likely, Clinton and her attorneys knew we didn't. We also felt confident that no Columbo-style "Aha!" moment would occur during the interview, when Clinton would inadvertently confess to some heinous crime. For one thing, she was too smart to be cornered into a confession of some sort, and her attorneys were too. But more to the point, we had yet to find any evidence that a crime had taken place, and we had looked. Hard.

Still, interviewing Clinton was a key part of the investigative process, and we were going to do it thoroughly and professionally. We prepped as much as we had for any interview we had ever done in Midyear. The two lead case agents — the two special agents on Wayne's squad with their names on the case file and the ultimate owners of the investigation — and analysts went over the interview outline and supporting exhibits again and again, conducting multiple dry runs with stand-ins for Clinton exhibiting varying degrees of cooperation.

While we prepared, the public attention to our investigation mounted, and so did the intense politicization. The NRA had begun running pro-Trump ads nationwide that focused on Clinton and Benghazi. We were painfully aware of the white-hot glare of Trump's campaign on the investigation and his public, preordained announcements of the outcome of our investigation.

As anticipation over the case's conclusion reached a fever pitch, logistics of the interview proved challenging for both our team and Clinton's. For Clinton, preparing for the questioning and then leaving the campaign trail for it constituted a significant amount of time and effort. For the investigators, finding a time and place to conduct the interview proved tricky. Because we would be discussing top secret information, the interview had to be conducted in a room approved for information classified at that level. That excluded hotel rooms and Clinton's residences in Chappaqua and Washington. In June we initially settled on using a suboffice of the FBI's New York City field office, but we later moved the interview to Washington. We settled on Saturday, July 2, a quiet holiday weekend when the Hoover Building would be deserted. We picked the windowless executive briefing room in the Strategic Information and Operations Center, the facility deep inside FBI headquarters where my team and I had been working for most of the past year. The interview room, which on a normal day would have been used for the president's daily brief to the attorney general and FBI director, was no more than a hundred feet from where we had run the investigation.

On the morning of the interview, the case agents and prosecutors who had worked with us from the beginning were organizing their notes when one of Clinton's attorneys called the DOJ attorney they had been dealing with during the investigation. We had arranged for the FBI police, uniformed officers who guard the Hoover Building every day around the clock, to let Clinton's car into the subbasement and to drop her off at a locked elevator bank that would take her directly to the floor just outside the interview room — a process sometimes used for VIPs who came into headquarters for sensitive meetings. Clinton's attorneys wanted to know if they could drive in separate cars, turning the quiet one- or two-car blip into a five-car motorcade. *I cannot believe this is the type of thing we're worrying about,* I thought with a silent groan. A polite no was the answer.

We moved our waiting game to the well-lit, antiseptic confines of SIOC, making final adjustments to binders full of tabbed information we would display during the interview. We soon got the call that the group had entered the building. Agents escorted them up through the garage to the basement elevator bank, then up to the locked outer door to SIOC, and finally to the interview room, with Clinton leading the way, smiling, energetic, and charming.

As she introduced herself, she spoke as if she had been briefed about the two case agents and myself, along with two of the attorneys on our side. Most people don't appreciate how much rehearsal — by both sides — goes into an interview like this. I had no doubt that she had spent days preparing. While I expected her to be ready for any question we threw at her, I didn't know how much her team would have prepped her about our team. By that time they had interacted extensively with all the agents and attorneys. They knew who was soft-spoken and who was brash, who turned on the charm and who was reserved. It made sense to me that they would have gamed out every aspect of the questioning, but it came across as effortless — the kind of easy familiarity that you'd encounter in any politician worth her salt.

A gaggle of Clinton's attorneys trailed in behind her. In addition to her lawyer David Kendall, she had two more attorneys from Williams & Connolly. Cheryl Mills and Heather Samuelson, her attorneys who had used the email sort laptops that had been such thorns in our side as we attempted to obtain and analyze them, completed the group.

The team and I had been irritated for days that Mills and Samuelson would be present. In addition to being attorneys, they were both fact witnesses, meaning that they had been present for or participated in matters we planned to discuss with Clinton. We worried that they might hear, interject, or advise Clinton during the interview about events they had been involved with themselves. I had raised this objection to their presence with the DOJ attorneys beforehand. The answer I received was that while it wasn't ideal, nothing under the law or bar ethics prevented them from attending — Clinton was free to choose whoever she wanted as her counsel. In fact, one of the DOJ attorneys had reached out to Kendall to convey our concern. After hearing him out, Kendall essentially responded, *Noted.* He was as aware as we were that Clinton could choose any attorney she wanted to be with her for a voluntary interview. Still, their presence seemed to me like another unforced error on the part of Clinton's team.

Like our earlier, months-long standoff about the sort laptops, Mills's and Samuelson's attendance at the Clinton interview seemed to me to be a tactical win but a strategic loss for their team. When Congress and the public inevitably found out about their attendance, it would be used as a cudgel to hammer home bogus arguments that the process was unfair and corrupt. What's more, neither of the two appeared to contribute anything during the course of the interview; indeed, with the exception of a few words from Kendall, neither Mills nor Samuelson nor any of the other attorneys spoke at all. Their presence just infuriated us as well as opened up later partisan criticism that Clinton wanted — and DOJ and the FBI allowed — witnesses to sit in on interviews of each other. Just as they had with the sort laptops, the Clinton camp seemed to be incurring self-inflicted wounds and making our jobs unnecessarily harder.

The government side of the table rivaled Clinton's. In addition to the two case agents who conducted the bulk of the interview, all four attorneys who had done the day-to-day work were present, as was the DOJ section chief in charge of the counterespionage section. While I understood everyone's desire to attend, there was no good reason that they all needed to be there. If we were going to add people, I would much rather have included an analyst or two who knew every last fact about Clinton, and I argued at length to try to reduce the number of people in the room. Panels don't conduct good interviews; the smaller the interview team, the more effective the interview. But my protestations were in vain, and I understood why. This wasn't a routine interview, and Clinton would not be lulled into letting her guard down, whether she was sitting across from two agents or eight of us.

We all sat down at the long table, with Clinton in the middle, directly across from the two case agents, Kendall to her immediate right and the other four attorneys to either side. I sat to the agents' left at the end of the long table, the four attorneys down the other end of the table, their supervisor sitting just behind us against the back wall.

Without small talk, the agents launched into the universal opening of every FBI interview: identifying themselves, thanking the interview subject for being present, telling her the purpose of the interview, and noting that it was a voluntary interview and she could stop whenever that she wanted to consult with counsel or take a break.

The content of the interview that followed is a matter of public record, written up in a form called an FD-302, which was released to Congress and to the public through several Freedom of Information Act requests by media and public interest organizations. The agents walked through an exhaustive outline, asking questions about everything from Clinton's use of email and her interaction with her staff, to her understanding of the email servers she used, to the individual instances of classified information she had received. To me, Secretary Clinton came across as intelligent, articulate, and well prepared. She was neither arrogant nor pandering. Not only did she answer the specific elements of our questions, but she also provided relevant context to the answer.

We were direct, but the atmosphere remained professional and smooth. That's not to say there weren't occasional problematic answers. For instance, when we showed Clinton an email that contained what is known as a "portion mark" — abbreviations such as (C), indicating information classified at the confidential level, that appear at the beginning of each paragraph of a classified document and note the classification level — she claimed she didn't know and could only speculate that the (C) was part of an alphabetized list of paragraphs. In other words, the former secretary of state claimed not to know the official notation for one of the three levels of classified information. I found that hard to believe, to say the least.

In another instance, we were trying to bore down into her understanding of what made any particular information classified. I stepped in and pushed the issue. I asked if she believed information should be classified if its unauthorized release could reasonably be expected to cause damage to national security, which is the statutory definition of confidential information. She replied, "Yes, that is the understanding." I thought to myself, *Well, that may be the understanding, but what's your understanding?* Here, and throughout the interview, our questions got quick but nevertheless carefully parsed answers.

Of course, choosing your words carefully is not illegal; it's what most well-prepared, intelligent people do. And hounding Clinton wasn't going to cause her to change her answers. But it was noticeable, and it was aggravating.

After three and a half hours we were done. Everyone shook hands, and Clinton headed back out on the campaign trail. After all the months of investigation, all the battles with her attorneys, and the crescendo of media and campaign focus on the investigation, the interview had been spectacularly anticlimactic. She answered the questions as we expected. And given how prepared she was, I'm betting that we asked the questions she expected. There was an element of Kabuki theater to our exchange: an enormously talented politician/former secretary of state/presidential candidate going up against a talented and well-prepared team of agents and attorneys. We largely had known what was going to unfold, barring any gaffes.

None of this meant that we weren't conducting a serious interview, or that we had prejudged anything. But everyone knew where the others stood, and we all had a shared knowledge of the facts of the case. Both sides understood that the investigation was at an end. If there were other investigative paths, we would already have pursued them. We all understood that we were unlikely to get any information beyond what we already knew.

In a predictable epilogue, Clinton's team promptly announced to the press that she had been interviewed in a "businesslike" and "civil" exchange during which she had answered all our questions, and expressed appreciation that they had been able to help bring "the review" to an end. I don't know if they knew — or cared — how much their refusal to use the word *investigation* bothered us. Needless to say, it didn't change what or how we did what we did, and I'm guessing their motives were entirely related to what some focus group had decided played best in the court of public opinion. But it seemed to me to be a microcosm of some of what the public detected in the campaign: a sometimes fuzzy relationship with the facts, and a failure to take responsibility for the behavior that had led to the investigation. *Just own it,* I thought, *don't weasel around it. Say, There's an investigation; we've done nothing wrong, and that will be proven out. Let's talk about something serious and not email.*

The fact that the interview was such a nonevent didn't seem to register with the partisan outrage machine, which ratcheted into high gear. By the early afternoon Trump had tweeted about Clinton and Midyear. "It is impossible for the FBI not to recommend criminal charges against Hillary Clinton," he wrote. "What she did was wrong!" Within hours of her interview, the *New York Times* had published a 1,000-word story about the interview packed with details supplied by sources who had either been in the room with us or who had spoken to people who had been there.

I was sadly unsurprised by the *Times* story. It was textbook public relations spin by someone who was clearly on Clinton's side: immediately and aggressively grab hold of the narrative, get a positive, simple message out, and demonstrate that the matter is over. I was confident that no one on our team was behind the leak, which logically benefited Clinton and her campaign, and I'm willing to bet that either her attorneys or political appointees at DOJ rather than anyone at the FBI had spoken to the *Times* or its sources for the article.

The interview was the capstone of the investigation, and it had only confirmed what Comey and the rest of the Midyear Exam team strongly suspected: that there was no case for charges. As the case agents retreated to computers in the Bubble to begin writing up the 302 (the standard form used by the Bureau to record the results of an interview, and the same FD-302 that would later be released to Congress and the public through FOIA requests), I gathered my notes — and my thoughts — for a call that was quickly approaching. Someone had set up a conference call for Comey; McCabe; James Baker, the FBI's general counsel; and the rest of the Midyear leadership team. I dialed in from SIOC and walked through the details of the interview, explaining what Clinton had told us and summarizing the unanimous belief of our team that we had not learned anything that changed our understanding of the facts of the case. There were a few questions, but not many.

With Clinton's interview finished, the stage was set for Comey's speech. He had seemed inclined to give one but had clearly wanted to wait until after her interview to make a final decision. Now it was over, and Clinton had not told us anything that changed our fundamental understanding of the gravamen of the allegations. The facts had settled right where we had anticipated, so Comey decided to move forward. Putting the final touches on his speech, he rehearsed throughout the weekend leading up to Tuesday, July 5, when he planned to hold a press conference.

We waited until the morning of July 5 to tell our counterparts at DOJ about Comey's speech. They were not pleased to have been left in the dark. Prosecutors from both DOJ headquarters and the U.S. Attorney's Office for the Eastern District of Virginia had been our partners from the earliest days of the investigation. Not telling them earlier about Comey's speech strained the trust between the FBI and DOJ teams. They knew the decision had been Comey's, not the investigative team's, and I hope they understood his reasoning, or at least came to understand it. In this early stage of summer, the investigation at least appeared to have ended with the correct legal conclusion.

The Bureau's Office of Public Affairs had invited the press to a large conference room on the ground floor. I wasn't sure we would be asked to attend, but that morning a handful of us were invited to sit in the back of the room, behind the press. No one paid any attention to us as we listened to Comey deliver the diligently crafted words that we had pored over so many times.

Hearing Comey talk about the investigation felt unreal. Midyear had been a large, stressful, and life-altering event, and Comey's speech provided closure. I thought about the team and the extraordinary work we had done on the investigation, how we had walked an arduous and difficult path and gotten it right. I wasn't thinking about whether or not making the speech was the right decision, how it would be portrayed in the presidential scrum, or what it might lead to. At the moment my pride was both a virtue and a fault, my admiration for the team and our work blinding me from seeing the ways it might — and later did — harm the Bureau.

Comey arrived at the end of his speech and announced his conclusions. Looking at the history of mishandling cases, we could not find a similar circumstance in which charges had been brought, he said. That didn't mean someone who did this would not face administrative consequences, and of course DOJ makes the final decision about prosecuting someone, but we believed no charges were appropriate. Comey concluded, "We did the investigation the right way. Only facts matter, and the FBI found them here in an entirely apolitical and professional way. I couldn't be prouder to be part of this organization." None of the team could.

As strobes flashed and shutters whirred, Comey collected his notes to walk out of the room. I glanced down from his lectern to the assembled media sitting between us. A reporter from a conservative network spun her head around, face contorted in surprise and rage. She seemed to be searching the faces in the room for a similar sentiment. To me, it was an ominous harbinger of things to come.

There was another bad omen on July 5, although I didn't know about it yet. On the same day that Comey spoke, an FBI source and former British government employee named Christopher Steele met with his Bureau handler and passed in an initial series of reports alleging contacts between the government of Russia and the Trump campaign. These documents, collected into a "dossier," would not arrive at headquarters for another two and a half months. When they did, they added to some of our worst suspicions about an entirely different, entirely more disturbing case—one that, at the time we began to close out Midyear, had not even gotten off the ground.

COINS

Finally, almost a year after we had entered the Bubble, Midyear Exam was complete. We briefed Attorney General Lynch at DOJ headquarters on July 6. After her infamous tarmac meeting with former president Clinton, Lynch had stated that she would accept whatever Comey recommended. Everyone had heard his speech the day before, so there was no doubt in our minds that she already knew what we were about to tell her. But presenting our recommendation to DOJ was a necessary formality, so we went through the motions all the same.

If Clinton's interview felt practiced, Lynch's briefing felt positively scripted. I filed into the attorney general's conference room with Comey; McCabe; Baker; Bill Priestap, the assistant director of the Counterintelligence Division; and Mary, the Midyear team's senior FBI attorney. Our prosecuting counterparts at DOJ were waiting for us. Above our heads, a gouge marred the wood paneling circling the vaulted ceilings — Baker told me it was unrepaired damage from a BB pellet that one of attorney general Bobby Kennedy's children had shot in the room. I sat down at the side of the room that looked out toward Constitution Avenue. Along the wall, portraits of former attorneys general stared down at us. One was a larger-than-life portrait of Elliot Richardson, Nixon's attorney general who resigned rather than fire special prosecutor Archibald Cox. I deeply admire him, and by fortunate chance I had met him through my wife just after I joined the Bureau in the mid-1990s.

We stood up as Lynch and deputy attorney general Sally Yates walked in with their staff. At the head of the table, with Yates to her right, Lynch listened as Comey began the briefing with a short introduction of the team. Then he handed off control of the briefing to the investigative leaders at DOJ, who went down the line of attorneys, each saying a brief word of praise or support of the outcome, before the baton was finally handed to the lower-level line attorneys with whom we had worked on a daily basis. They walked through an analysis of the charges that might apply to Clinton's case and the various flaws with the application of each, spending considerable time discussing gross negligence, how that had been variously defined, and DOJ's past application of the statute containing the gross negligence provision. They landed in the same spot as Comey: no reasonable prosecutor would bring charges with the facts we had uncovered.

Yates asked a few questions. David Margolis, the senior attorney at DOJ who was not an appointed official, weighed in as well. A man whom the *Washington Post* has described as an "all-knowing, Yoda-like figure," Margolis had begun working for DOJ in the Johnson administration. He had guided DOJ through political storms for 40 years. Recently back to work after medical leave, he listened carefully from his seat at the opposite end of the table from Lynch, then pronounced, like Solomon, *If we were to pros-ecute under these facts, it would amount to celebrity hunting. And we don't do that.*

Lynch took all this in. Then she turned toward Comey and said she ac-cepted his recommendation. She thanked us, and asked that we convey our thanks to the team for their work. She asked her staff to prepare a public statement announcing her decision, which would be issued with lightning speed, before I sat down to dinner that night.

After an effusive round of self-congratulatory thank-yous, we stood to leave. The Clinton investigation was done, we believed. We were free.

As we left the room, I found myself walking behind Margolis. While I had never met him, I respected his reputation and his experience of work-ing through the tumult of the civil rights movement, Watergate and Nixon, and other challenges. I thought fleetingly about introducing myself and thanking him, but I decided against it. It would have felt unprofessional, more like a starstruck fan approaching a celebrity than one governmental official engaging with another. Looking back, I was probably wrong, and I'm certain he wouldn't have minded me introducing myself, but we left the room without exchanging a word. He passed away nine days later, having given his wisdom and imprimatur to one last chapter of department history.

Although we were done with Midyear, Trump wasn't. Chants of "Lock her up!" continued at his rallies, and he continually questioned our moti-vations and competence. Not me or the team, but the entirety of the FBI.

The political stigma attached to the investigation bothers me to this day. As a young agent in Boston, I once heard an agent favorably described as someone you would want as the case agent if your child was kidnapped. That characterization of the height of praise remains in my head. The Mid-year investigative team was just such a team: professional, discreet, curious, relentless. Solid across every aspect of what we did. I would not have chosen anyone different. But the respect and prestige due to the team members from working on such a complex, challenging, and high-stakes investiga-tion were eclipsed by the political debate about its process and outcome. Whatever your opinion of Comey's decision to speak about the results of the case or the decision not to prosecute, know this: the investigation was conducted by an extraordinary team that their peers and the American people should be proud of.

The FBI and DOJ both award public honors for investigative achievement. In an ordinary case, the Midyear team's work would have made it a frontrunner for both the director's and the attorney general's awards. But the political reality made it too sensitive for such high-profile recognition. Comey knew this. Toward the end of the investigation, his chief of staff asked Derek and me to count up the number of people on the core Midyear team. Like the challenge coins that military commanders and law enforcement authorities give out to people to commemorate a unit or some noteworthy endeavor, Comey wanted to make Midyear coins that he could present privately to each team member. We did the head count and gave him the number, after which he designed a coin and paid for them himself.

But by the time the coins arrived, the political vortex around the investigation had gained strength. Comey's chief of staff and I agreed to wait until the political climate cooled. We thought that would take only a few months. To this day, the Midyear team is still waiting for the coins.

6

Crossfire Hurricane

ON THE EVENING of August 1, 2016, I waited impatiently in the clogged aisle of a crowded Airbus A380 parked at the terminal at Dulles Airport, inching through coach as the British Airways passengers in front of me took their seats. We were boarding a redeye flight that would carry us across the Atlantic to a major European city, a capital whose name has not been acknowledged by the FBI as of this writing and which I therefore will refrain from identifying here. I carried only a small go bag over my shoulder containing a navy-blue suit, two white shirts, a few ties, and other clothes for a short stay. A pocket held a handful of foreign currency scrounged from a loose-coin box from previous visits.

The slow-moving conga line of passengers shuffled into the belly of the plane. Somewhere else on the enormous aircraft, an FBI agent named Luke — a Washington field office supervisor with extensive experience with Russia and its intelligence services — jostled through the aisles toward his own seat; when traveling by plane, agents usually sit separately, to minimize attention.

When I reached my seat, I buckled myself in after stowing my bag, which contained nothing that would identify me as a counterintelligence agent. No briefing documents, no Bureau laptop, no encrypted thumb drives. Nothing that might raise questions.

I pulled out the safety card and reflexively glanced around the plane: 16

emergency exits for the almost 500 passengers aboard. As I settled in for the long trip, I recalled a long-ago flight in the summer of 1973. I was not even four years old and was traveling back from the U.S. with my parents to our home in Tehran. At that time the airline was known as the British Overseas Airways Corporation, or BOAC. It was neither reliable nor luxurious in those days. British expatriates called it "Better On A Camel."

Conditions had improved on transatlantic flights in the intervening four and a half decades, and British Airways always had good in-flight movies and entertainment. This wasn't just an opportunity to catch up on pop culture. Earphones and a movie screen were convenient buffers from prying fellow travelers, who might ask about my business abroad. If I had to respond, I had stock answers. *I'm on vacation,* I might say. Or, *I'm visiting relatives.* Never, *I'm traveling on business,* which begged follow-up questions about what line of work took me to Europe.

The reason was obvious. I could never say I worked for the FBI, let alone reveal that I was investigating what might be the most consequential and chilling case of my career: whether a major political campaign had conspired with a foreign enemy to throw the outcome of a presidential election.

This new investigation was a nascent one — a mere thread compared to the fully formed tapestry of the Clinton case that we had just wrapped up. But its political and diplomatic implications were so profound, so potentially explosive, that our trip across the Atlantic had been organized under utmost security. Only a tiny number of high-level Bureau officials knew about it. In time it would be the case that changed my world and our country.

Numb with fatigue, I looked out the window as the plane lumbered down the runway. A low ceiling of clouds made an early dusk for the hazy August evening. After wheels-up, the plane banked slowly eastward toward the Atlantic and Europe. When the flight attendant came through the cabin, I ordered a scotch and soda. Since I was traveling without a firearm — very few nations allow visiting agents to bring weapons — FBI rules permitted me to have a drink. I eased back my seat into a reclining position. I didn't expect to talk much at all on the long flight. I would need plenty of rest.

"RUSSIA, IF YOU'RE LISTENING"

A head-spinning several days had preceded the rushed preparations for the trip. After Midyear Exam concluded with Comey's speech in early July, we barely had a chance to get our bearings, to take a week off to breathe, when a communication arrived from overseas that shook me to my core — and that led directly to the opening of a new investigation, this one into people associated with Hillary Clinton's main rival for the office of the presidency.

The communication, which came by way of our office in the European capital to which I was now headed and had been sent by one of America's international allies, had been precipitated by a public statement by Donald Trump three weeks after Comey's speech announcing the end of Midyear Exam. On July 27, candidate Trump had held a press conference in Doral, Florida, where he made an astounding appeal through the assembled members of the media: "Russia, if you're listening, I hope you're able to find the 30,000 emails that are missing," he said, referring to the email that Clinton's team had determined to be non-work-related, which we had found no evidence to indicate she had attempted to hide. "I think you will probably be rewarded mightily by our press."

The statement, which Trump subsequently characterized as sarcasm, stunned and alarmed intelligence officials and foreign policy experts, not to mention ordinary American citizens and international observers who respect the democratic process. A major candidate for the presidency of the United States had invited a foreign adversary to meddle in an election. It was a watershed moment — the first of many to come.

Within five hours of Trump's speech, the Russians launched a new cyberattack on a server used by Hillary Clinton's personal office. Unit 26165 of the GRU — the same military intelligence unit that had attacked the DNC and the Clinton campaign in the spring — fired off a fresh round of spearphishing email containing malicious links to 15 email accounts connected to the Clinton campaign. Since March the GRU had launched numerous spearphishing attacks, including one against the Democratic Campaign Congressional Committee, which opened the door to virtually the entire national Democratic Party computer network. Between April and early June, Unit 26165 had successfully tunneled into more than 30 computers on the DNC network, its mail server and shared file server, and had begun releasing the stolen information in mid-June. Now Trump appeared to be asking for more of the same, targeted at his political adversary—and the Russians appeared to be delivering.

Trump's exhortation to Russia rocketed through the media domestically as well as overseas. One of the people listening was a trusted foreign diplomat who later publicly revealed himself to be Australian high commissioner Alexander Downer, Australia's ambassador to the United Kingdom. In Downer's recounting, Trump's words jarred his memory of a series of conversations months earlier. In early spring of 2016, Downer and a colleague were at an upscale London bar to meet a foreign policy adviser to Donald Trump. The adviser was George Papadopoulos, the energy consultant who was one of the few volunteers to fill an embarrassingly empty slot in the Trump campaign and who had been working eagerly to establish a rapport with the Russian government.

During conversation with the Australians that spring, Downer said, Papadopoulos had made a jaw-dropping comment: that the Russians had obtained damaging information about Clinton and President Obama. And, Papadopoulos continued, the Russian government had offered to assist the Trump campaign through a coordinated release of the material. Although we didn't know it at the time, it was the same information that Papadopoulos had received from Joseph Mifsud.

At this point I've reached an awkward juncture in the narrative, a place where I am prevented from saying more. The Mueller Report states that Papadopoulos made the statement on May 6 but says only that Papadopoulos made the statement to "a representative of a foreign government." A report issued in late 2019 by the DOJ inspector general goes slightly further, calling it a "friendly foreign government," or FFG. Beyond that and the public statements of Downer and Papadopoulos, nothing has been declassified, and I cannot provide more detail than I have here. Nor, in fact, can I even confirm whether Downer's and Papadopoulos's statements are true. Because of those restrictions, I must adopt the somewhat stilted descriptors that follow.

What I can say unreservedly is that Papadopoulos's information was alarming to us on several levels when we learned of it on July 28. It came to us in the form of an email from our legal attaché (or LEGAT, in Bureau-speak, the senior FBI agent assigned to a U.S. embassy) in a European capital; this American official had forwarded an excerpt of a report about the meeting from the friendly foreign government. As I reviewed the information that the FFG had shared, my heart sank. If this information was true, it signaled a dramatic escalation in Russia's shadow war against American democracy. It also would indicate that Moscow believed it could count on some Americans to further its efforts — and that this belief was not entirely unfounded.

First, Papadopoulos's admission predated the release of the stolen Democratic email on the website DCLeaks, and Julian Assange and WikiLeaks hadn't yet made comments about the future release of Clinton's email. None of the information about Russia's cybertheft was in the public domain. In other words, Papadopoulos had somehow learned about the hacking operation before the public did and had advance knowledge of the Russian plan to use that information to hurt Clinton's campaign. Even the FBI hadn't known about it at that time.

Second, the intelligence in the email from the LEGAT wasn't simply that the Russians had the harmful information. According to Papadopoulos, they had offered to coordinate its release to help the Trump campaign. Beyond a simple cybertheft, Russia was offering to conspire with the campaign to release the stolen information in a way optimized to help the Trump campaign. It was classic Russian *kompromat* but being used in a new and disturbing way: not to manipulate an individual into taking actions that would benefit Moscow, but rather to manipulate — and undermine — an entire nation's political system.

The email from the LEGAT arrived like a thunderclap out of the blue. When we received the report about Papadopoulos's revelations to the FFG's personnel — intelligence that they then sent from their embassy to ours shortly after Trump's Florida press conference — we already had a broad idea of some of what the Russians were doing to influence our elections, particularly including cyberattacks. We knew, for example, of the successful intrusions against the Democratic National Committee and the Democratic Congressional Campaign Committee. We knew that Guccifer 2.0, whom we believed — and later confirmed — to be a front for Russian intelligence, had begun strategically releasing those stolen email messages. We also had initial hints that Russian cyberwarriors had begun probing state-level electoral systems.

The allegations about Papadopoulos changed our understanding of the scope of what the Russians were willing to do. Until the report arrived from our LEGAT, we had not seen evidence that the Russian efforts extended to engaging with a presidential campaign or meddling directly in an election. Nor had we yet grasped the Russian threat in its entirety: a broad, multifaceted attack on our electoral process. To say this was a radical increase in Russian aggression would be an understatement.

As we began to understand the breadth and depth of the Russian activity, much of which still remains classified, we understood that we were looking at a clear and widespread assault on our electoral system, from exploiting our freedoms of speech and the press to undermining the public's faith in our political system. Now we had to figure out what the Trump campaign knew about this bold Russian campaign, a campaign that, as was becoming increasingly clear, was designed to help Trump. And — although none of us relished this possibility — we needed to know if the candidate or his associates had participated or were participating in this attack on the very democracy they sought to lead.

INSOMNIA

Late in the evening of Thursday, July 28, I sat in Bill Priestap's office listening to Derek tick through the initial intelligence about Papadopoulos and the Russian efforts. I had known Bill, a tall, lean, and quietly thoughtful man, since 1998, when we were Quantico classmates. People occasionally misjudged Bill's earnestness as a sort of deferential intellectualism. That was a mistake. Bill's Michigan roots imparted a quiet modesty that prevented him from talking about things like his law degree or pre-Bureau employment as an assistant coach on the University of Michigan football team, and hid a sometimes stubborn determination once he made up his mind about something. Derek was the man I had had by my side for nearly every day of the past year on Midyear Exam: logical, fact-based, precise. For someone who was almost completely unflappable, he spoke with an understated concern that conveyed how bad things might be.

We had never heard of Papadopoulos before receiving the FFG report. We quickly determined that he had a legitimate connection to the Trump campaign. On March 21, Trump had listed his foreign policy team in an interview with the *Washington Post*, describing Papadopoulos as "an energy and oil consultant, excellent guy," in the same breath with Carter Page, Walid Phares, Joe Schmitz, and retired lieutenant general Keith Kellogg.

Several aspects of the initial reporting about Papadopoulos have been difficult to explain to those outside counterintelligence circles. Most people aren't aware of how the FBI collects information and vets it, how investigators determine what is accurate and what isn't, and why people give the FBI intelligence in the first place. Understanding the process matters, because the confusion around the FBI's procedures and techniques has allowed partisans, right-wing media outlets, and conspiracy theorists to cast doubt on the tip that launched the FBI investigation into the Trump campaign — an investigation that was built upon solid intelligence, something you might not be aware of if you subsist on certain forms of cable news and social media.

First, the source of this information was highly reliable — and not at all unusual in the FBI's or any other intelligence agency's counterintelligence work. Some members of Congress have groused that the information was somehow improper or invalid because it hadn't come through "intelligence channels." That claim reflects either woeful or willful ignorance, and possibly both. The FBI can and does gather credible allegations from overseas, including counterintelligence information. The information can come from an intelligence agency, from a newspaper, or from an alert businessman, a suspicious scientist, your neighbor, or your grandmother. It certainly can come from a trusted foreign diplomat.

Second, the FBI's response to this information had everything to do with the Russian threat and nothing to do with the Trump campaign. The urgency wouldn't have been any less if Papadopoulos had worked for the Clinton campaign, or the campaigns of Sanders, Stein, Cruz, Kasich, or anyone else. The attack — and it unquestionably was an attack — was an intelligence operation against the United States and our presidential election, a sacred pillar of our democratically elected governments. As the counterintelligence wing of the FBI, we were mandated to take this attack seriously and to confront it head-on, no matter where our investigation took us.

Third, the Bureau's initial response to this information was to scrutinize it to assess its reliability — not to weaponize it, as many have claimed. All of us approached the Papadopoulos information, like every allegation, with a healthy degree of skepticism. No good investigator takes information at face value. If anything, the opposite was true: I knew something was being advanced as a fact, but we needed to verify it. Preferably through several independent ways. This skepticism constitutes a fundamental reason that the FBI conducts investigations. Indeed, the need to find the truth put me on that redeye to the European capital. Despite our natural skepticism, this information was significant, and we had to determine its veracity.

A fourth point rattled through my head over and over after I received the tip from the FFG: this wasn't normal. It was like a jingle I couldn't get out of my head. None of us had seen anything of this nature before. It was unheard of, and unlike anything we had seen in any prior administration. Its gravity and potential for sinister consequences would disturb any intelligence professional. It wasn't like most allegations involving political actors I had seen over my career. It wasn't an agent three layers removed from a foreign government trying to cozy up to a sibling of a candidate in a business deal. It wasn't someone with family living overseas in a hostile country trying to gain employment on the local staff of a cabinet member. This was a leading political candidate for the most powerful position in the United States and arguably the world.

Counterintelligence work frequently feels like studying a placid pond, searching for the slightest ripple on the surface. This was like a 300-pound cannonball screaming into the still water.

Long after dark, Derek's briefing ended and I left the office. As I departed from the building, Washington was quiet and still. I drove home feeling deeply unsettled. It took me a long time to fall asleep.

I awoke at 4 a.m. and lay with eyes open in the dark, my thoughts tumbling over the Papadopoulos information and what it meant. Why and how would someone as young and inexperienced as Papadopoulos have heard this information? Was this a planned Russian intelligence operation, or something less formal and more opportunistic — just a shadowy information broker trying to connect people and governments to burnish his reputation and earn IOUs? Or was this all junk, simple boasting or disinformation seeded by the Russians to chew up our resources trying to figure it out? During the Midyear investigation I had experienced occasional sleeplessness. That insomnia was nothing compared to what I was experiencing now, though — much less what was to come.

Nearly every morning after that first briefing, I would jolt awake at the same time, my thoughts racing over the investigation ahead and its implications. What we knew was less concerning than what we didn't, and I was constantly preoccupied with thinking about ways we could find information to fill the gaps in our knowledge. Above all, I was aware of the unstoppable march of time. The election was only three months away. If there was a problem, we had almost no time to figure it out.

As it turned out, of course, this new investigation wouldn't end in November 2016. It would be more than three years until I would sleep well again.

UNSUB

The sun finally rose on Friday, July 29, and a dreary morning greeted me as I sped downtown to my office. An email was waiting for me with everything we knew so far. The report from the FFG was detailed and credible, but there were gaps. We couldn't yet identify whether the Russian overture was initially made to Papadopoulos or to someone else. Nor was it clear whether the Russian offer had been pursued.

Later that morning, following the daily intelligence briefings between the deputy director and his senior staff, Bill briefed McCabe. Although the allegation involved a likely Russian cyberattack, due to its nexus to Russian intelligence activity, McCabe wanted the Counterintelligence Division rather than the FBI's Cyber Division to run the investigation. And he wanted the investigation opened immediately. He conveyed a sense of urgency: our investigative work was to be measured in hours and days, not weeks and months. Bill conveyed the order: *Open the case now.*

The first step was to learn anything more that we could about the allegation. I was told to attempt to interview people in the European capital who could provide us with further information — although again, I can't say who. We also knew we had to reach out to trusted members of the U.S. intelligence community to ask, delicately, what additional information they might have. That meant going to a select group of key members, but doing so with the utmost discretion.

Midyear had publicly thrust the FBI into the middle of a campaign, an extremely uncomfortable and unwanted position. The last thing we wanted was to be back in that position again. Moreover, on the heels of that investigation, we worried that leaks about this new matter could get out and hurt Trump, affecting the general election. Clinton's email scandal and the FBI investigation that followed had been the subject of intense public interest from day one. This was different: no one outside the Bureau knew of the allegations about Trump's campaign or was aware that we were investigating. And we aimed to keep it that way. The sensitivity of the case warranted extreme layers of secrecy, and we began what would be a ubiquitous aspect of the investigation: bending over backward to make sure no word escaped into the public domain, even if it hurt our efficiency.

To keep the matter as compartmented as possible, McCabe decided to talk to an extremely limited number of his counterparts in the intelligence community. The agencies with which he spoke were our closest partners, and in the world of counterintelligence, because of their size and access to information, they, along with the FBI, are the heavyweights in the room. To dispel other rumors that have swirled around Crossfire Hurricane, no information from any intelligence community member played a role in the opening of the case. The report from our LEGAT provided the sole basis. Period. But while we started the case on our own, we needed to coordinate to finish it. To get to the bottom of the allegations, particularly given the small window of time we had, we needed help.

As with our most sensitive investigations, we carefully limited access to our work, setting up a tightly controlled list of individuals with knowledge of the case — known as a "bigot list" in intelligence parlance, a term dating to World War II that denotes the highest level of secrecy — and prohibiting access to the investigation for anyone else in our file and data systems. Essentially, we made it impossible for anyone other than the small group of investigators assigned to the case to learn anything about our investigation, much less see the information it was generating.

The public is often frustrated by the FBI's silence about ongoing cases and the standard "neither confirm nor deny" response to requests for comment. But important considerations inform keeping investigations secret and not discussing them publicly. Protecting the reputations of innocent parties is key. Sometimes the FBI investigates people who don't end up being guilty. Sometimes it opens cases on groups of people knowing that the investigators are looking for only one person among them. For all we knew, at this point in this particular case, Trump was a bystander to a misdeed that Papadopoulos, and Papadopoulos alone, had perpetrated.

Another reason for secrecy in the FBI's counterintelligence work is the fundamentally clandestine nature of what it is investigating. Like my work on the illegals in Boston, counterintelligence work frequently has nothing to do with criminal behavior. An espionage investigation, as the Bureau defines it, involves an alleged violation of law. But pure counterintelligence work is often removed from proving that a crime took place and identifying the perpetrator. It's gaining an understanding of what a foreign intelligence service is doing, who it targets, the methods it uses, and what the national security implications are.

Making those cases even more complicated, agents often don't even know the subject of a counterintelligence investigation. They have a term for that: an unknown subject, or UNSUB, which they use when an activity is known but the specific person conducting that activity is not — for instance, when they are aware that Russia is working to undermine our electoral system in concert with a presidential campaign but don't know exactly *who* at that campaign Russia might be coordinating with or how many people might be involved.

To understand the challenges of an UNSUB case, consider the following three hypothetical scenarios. In one, a Russian source tells his American handler that, while out drinking at an SVR reunion, he learned that a colleague had just been promoted after a breakthrough recruitment of an American intelligence officer in Bangkok. We don't know the identity of the recruited American — he or she is an UNSUB. A second scenario: a man and a woman out for a morning run in Washington see a figure toss a package over the fence of the Russian embassy and speed off in a four-door maroon sedan. An UNSUB.

Or consider this third scenario: a young foreign policy adviser to an American presidential campaign boasts to one of our allies that the Russians have offered to help his candidate by releasing damaging information about that candidate's chief political rival. Who actually received the offer of assistance from the Russians? An UNSUB.

The typical approach to investigating UNSUB cases is to open a case into the broad allegation, an umbrella investigation that encompasses everything the FBI knows. The key to UNSUB investigations is to first build a reliable matrix of every element known about the allegation and then identify the universe of individuals who could fit that matrix. That may sound cut-and-dried, but make no mistake: while the methodology is straightforward, it's rarely easy to identify the UNSUB.

Take the imaginary recruitment in Bangkok. Does "American intelligence officer" mean what we think it means? Could it be a defense attaché? Or maybe they were wrong about its being an intelligence officer and it was actually a State Department official? Someone passing through on a temporary duty assignment? Any one of them could be the UNSUB.

And what do we actually know about the source in Moscow—the Russian whose tip caused us to open the hypothetical Bangkok UNSUB case in the first place? Sources are problematic, sometimes imprecise, usually complicated. They embellish. They forget things. They make things up. They believe their memory is more reliable than it is. They have drinking problems. Marital problems. Gambling problems. They want to be liked, admired, paid, to have purpose. Everybody's looking for something, as the Eurythmics said.

Investigators try to identify and understand these factors. Most of all, they ask questions with analytic rigor. Does the source have a record of reporting reliable information? Was he groomed and recruited, or was he a walk-in who volunteered to us, which raises the risk that he's part of a Russian operation to feed false information? Do we know what various components of the U.S. intelligence community know about the source? What if a different source conveys a similar story from the same time frame, except that that source heard the recruitment was in Kuala Lumpur? Is that the same allegation or a different one? What if the Russians recruited *two* people? Most everyone had been drinking at the reunion, so how much of this is solid information to begin with?

Some of the most confounding real-life—and still unsolved—cases involve multiple sources, where a more reliable source might have vague information while less vetted sources provide conflicting but very specific information about the same allegation. The variables can be limitless, and involve constant reassessment and balancing of source information. Intelligence work, and counterintelligence in particular, is sometimes called a wilderness of mirrors. It's hard enough to figure out the truth, but when your opponent is actively trying to muddy your view of reality, it can be diabolically complex (and as addictive as the best puzzle you'll ever encounter).

The FFG information about Papadopoulos presented us with a textbook UNSUB case. Who received the alleged offer of assistance from the Russians? Was it Papadopoulos? Perhaps, but not necessarily. We didn't know about his contacts with Mifsud at the time — all we knew was that he had told the allied government that the Russians had dirt on Clinton and Obama and that they wanted to release it in a way that would help Trump.

So how did we determine who else needed to go into our matrix? And what did we know about the various sources of the information? Papadopoulos had allegedly stated it, but it was relayed by a third party. What did we know about both of them: their motivations, for instance, or the quality of their memories? What were the other ways we could determine whether the allegation was true?

And if it was true, how did we get to the bottom of it?

One thing was clear: to answer these questions, I needed to go to the European capital from which this information had been sent.

JUMPIN' JACK FLASH

It was midday Friday when I picked up my office phone in Washington — late afternoon in Europe. I was calling an old friend of mine in the FBI office at the U.S. embassy there to see who would talk to us — those parties the government has yet to name. I wasn't hopeful, given the explosive nature of the information we'd received and most nations' reluctance to involve themselves in an ally's internal political affairs. Still, I needed to ask.

At headquarters we had previously kicked around the idea of doing the interviews over the phone, which I wasn't in favor of. In my experience, and that of most investigators, the best information comes from face-to-face meetings. No phone call or email or videoconference can substitute for an in-person, direct discussion. Fortunately, our LEGAT agreed and, based on a signal from the FFG that they might be willing to talk to us, felt the best option was to try and meet them in person. As a result, I was calling my colleague to put a plan together.

The friend in the European capital was an assistant legal attaché (or ALAT) with a caustic sense of humor and a deep counterintelligence background. He also had the charm to implore, cajole, and otherwise expedite my request for a meeting with our foreign counterparts. It was midsummer; many people in the U.S. embassy were on vacation, and an acting deputy chief of mission was in charge. The good news was that the acting deputy was a career diplomat and a professional. The bad news was that he might want an additional check from Foggy Bottom in Washington before signing off on our request to travel. When conducting an investigation abroad, the FBI needs a variety of approvals, including in this case, from the people we wanted to talk to. And we needed to get it as quickly and quietly as possible.

Our speed also complicated the process. It's the nature of bureaucracy to slow action and mitigate risk, to ensure that laws and regulations are followed. The harder the decision, the higher the level of approval required. And then people are busy, or lazy, or just aren't particularly good at their job. An action that isn't normal, a task or approval that doesn't fit the typical routine, can mean a death knell for expediency. And this trip was far from normal.

I also needed an experienced investigator with me on the trip. After checking first with senior management in the Washington field office, I used a secure line to call the subordinate supervisor I hoped to take with me — and ultimately to bring to headquarters to supervise the investigation. It made sense to initially conduct the investigation from headquarters because of its extreme sensitivity and until we determined which field office the subject lived near, at which point we could send the investigation there. It wasn't just the supervisor's Russia experience that made him an excellent fit for that leadership role. Some of the best agents I know succeed because of their tenacity. Curiosity and intelligence are important, as are judgment and perspective. But a relentless drive and a dogged focus on the mission were the traits I looked for first in agents.

I quickly briefed him on the facts and asked him to get a bag ready to go to Europe to do some interviews.

When are we leaving? he asked me.

No idea, I told him. *Probably not until Monday, but I want to be ready to go tomorrow.*

How long are we going for? he asked.

I don't know, I admitted. *A few days at most.* I wasn't sure if we would get to yes with our counterparts, but our sitting there in Europe would make it harder for them to say no.

I had work to do before we could depart. When I left the office on Friday, I grabbed my assigned take-home laptop, configured to operate at a classified level on our secure network.

McCabe's directions had been clear — the Papadopoulos intelligence was an urgent priority, and the FBI counterintelligence team needed to open an investigation immediately — and Priestap had authorized opening the investigation. But with the information just days old, the actual administrative act of opening the case hadn't been done yet; in the swirl of events, there simply hadn't been time to put together all the paperwork making official an investigation that in effect had already been set in motion by McCabe's directive. As we made plans for our travel and interviews the next Monday, agents and analysts had already begun diving into the investigation. While it isn't improper or unusual for the paperwork opening an urgent case to trail the beginning of investigative work by a day or two, I didn't want to return from overseas to find out the case had still technically not been opened.

It was unusual to open a case from home on a weekend, but the urgency of the situation gave us no other choice. Over the weekend I talked from home with Derek and attorneys from the Bureau's Office of General Counsel about the wording of the opening document, which sets forth the justification and purpose of the investigation. This document would be the foundation of the case, the first page of the first chapter of everything that would unfold later, and I knew it would inevitably be scrutinized in the future. We needed to make sure that every aspect of the case opening was done by the book and that the document included all the required elements in our investigative guidelines.

Sitting in my home office, I opened the work laptop and powered it up. The laptops were balky and wildly overpriced, requiring an arcane multi-step process to connect. They constantly dropped their secure connections. Throughout the D.C. suburbs, FBI agents flew into rages when the laptops quit cold while they were trying to work at home. Chinese or Russian intelligence would have been hard-pressed to develop a more infuriating product. Nevertheless, they let you work away from the office.

After logging in, I pulled up a browser and launched Sentinel, our electronic case file system. Selecting the macro for opening an investigation, I filled in the various fields until I reached the blank box for the case name. I paused to think. When opening a new case file, agents may opt to use an automated computer program that spits out random two-word combinations to select a name for the case. There's a second, low-tech method: the agent can choose a name, provided the codename hasn't been used before.

Codenames serve as the title for any investigation, both for filing purposes and as shorthand for referring to the investigation in conversation. The names of espionage cases and other sensitive counterintelligence investigations are carefully selected to conceal the identity of those under investigation. The rationale is straightforward. The name serves as an additional layer of secrecy to mask the subject of the investigation. The cardinal rule is that the name shouldn't allow an identification of the subject or anything about them — that would defeat the whole purpose.

Imagine that the computer naming program generates the codename Common Disaster for the hypothetical example of our spy in Bangkok. Common Disaster is a lot to write, so agents will inevitably refer to it as CD when talking or emailing. As we gather information about people we're identifying under the CD umbrella, we put it into the matrix, essentially a big grid: on one axis is a list of all the people we know, and on the other a list of the data points in the allegation: whether the person is an intelligence officer; whether he or she was in Bangkok at the time in question; whether he or she had prior contact or CI issues with the Russians or other security-related incidents or concerns. As our understanding of the allegation evolves, so do the elements of the matrix.

Using two words also has an ancillary benefit: it easily allows for the naming of spinoff investigations. If three suspects emerge in the matrix of the Common Disaster investigation, they get their own investigations, with linked codenames — Common Shade, Common Pill, and Common Grifter, for example. Or, because we hate to type, CS, CP, and CG.

As I sat at my desk staring at the insistently blinking cursor, I racked my brain for a good codename. The words of the Rolling Stones song "Jumpin' Jack Flash" ran through my mind, Mick Jagger's swaggering "I was born in a crossfire hurricane / And I howled at the morning driving rain." Then I typed in the codename that would follow the case forever: Crossfire Hurricane. I didn't appreciate how prescient that title would be.

BRASS IN POCKET

That Sunday night I laid out my suit and shirts, carefully folded them, and rummaged through a small jewelry box looking for the right foreign currency. For once, I remembered to toss in an extra power adapter; I have an entire drawer filled with overpriced converters purchased at airports and hotel front desks around the world.

The morning of Monday, August 1, brought the good news that the people in the European capital whom we wanted to talk to appeared willing to work with us, though everyone remained concerned about the potential political ramifications. The more detail we could get about what Papadopoulos had said to the allied nation, the better chance we had at figuring out who was the UNSUB at the heart of Crossfire Hurricane. I was encouraged, though wary of getting my hopes up lest I jinx the effort.

Looking back now, I am bemused by the fact that at the time we thought we would eliminate potential subjects until we arrived at a one-person, one-outcome case, rather than the multiheaded hydra of an investigation that would emerge in the coming weeks. In hindsight, some people seem to have been less optimistic than others. Shortly after we opened Crossfire, CIA director John Brennan approached Comey about disturbing intelligence that was beginning to come in about Russia's burgeoning interference in the elections. Because of its international and domestic scope, getting to the bottom of it was bigger than any one member of the intelligence community could handle, and the two agreed to set up a small interagency team to get our heads around what the Russians were doing. Though the case had only just begun, the heads of the CIA and FBI thereby set up a scaffolding for a related, broader effort — one that would involve the wider U.S. intelligence community in the urgent task of deciphering what was going on in Russia, which cyber actors and activities were being deployed against the U.S., and what intelligence actions the Russians were engaging in.

Crossfire would be part of this interagency effort. While all the members focused on what the Russians were doing, the agreed-to caveat between the agencies was that the FBI would wall off anything relating to our investigations of U.S. citizens from the rest of the intelligence community, as we were legally required to do. Papadopoulos was an American, so the FBI alone was going to Europe, and quietly.

With the trip now a certainty, I cleared out my email box and turned on the out-of-office autoreply. Then I logged onto our travel site, bought an airline ticket, and made my way out to Dulles to meet up with my new partner, Luke.

7

Among Friends

IN SPITE OF my fatigue, I slept at most 90 minutes during the six-hour flight. As we landed in Europe, rain streaked the Airbus windows. Weary from the whirlwind earlier efforts and now from jet lag, Luke and I made our way through Tuesday morning traffic to the U.S. embassy. We met our local FBI colleague and walked to Post One, the shared title of the primary security desk in every embassy, for access badges.

On our way to the FBI offices with our new badges, we stepped into a cramped bathroom to clean up. We balanced our opened, half-unpacked luggage precariously on the windowsills as we shaved and changed from our travel clothes into our suits. Standing side-by-side in front of the mirrors, I noticed Luke carefully adjusting a white dash of fabric in the breast pocket of his suit.

Are you wearing a fucking pocket square? I asked, eyebrow raised.

Dude, he said, smiling. *We're in [the name of the European capital]. Besides, this looks good.*

He was right, I thought grudgingly. It did look good. And I would never in a million years admit it. *Come on, GQ,* I said, *let's go meet the boss.*

Every U.S. embassy's senior leadership is known as "the country team." In my experience, country teams were like Tolstoy's families: the happy ones all functioned the same, while troubled ones had their own unique dysfunction. Fortunately, this embassy's country team was one of the happy ones. From a counterintelligence perspective, the most important senior players in some embassies include the ambassador, the deputy chief of mission, the CIA chief of station, the FBI's LEGAT, and sometimes the State Department's regional security officer. Work that touches on other FBI investigations might include representatives from the Drug Enforcement Agency or the Departments of Defense and Homeland Security.

All of those entities have overlapping responsibilities. The ambassador is always the president's highest-ranking emissary in any country. While it may sound like an overstatement, he or she has the final say on almost anything that occurs within the country related to the United States. The CIA chief of station typically represents the director of national intelligence and as such oversees all intelligence activity, including that of agencies other than the CIA. The regional security officer from the State Department's Bureau of Diplomatic Security is responsible for embassy staff safety, for the embassy's counterintelligence posture — including overseeing locally hired foreign service nationals, who are indispensable but some of whom pose serious intelligence vulnerabilities — and for the physical security of the embassy itself. The FBI investigates espionage committed by U.S. government personnel assigned to embassies and consulates abroad. And, crucial to the Russia election investigations, the FBI participates in counterintelligence activity that connects to its domestic mission.

While those overlapping responsibilities sometimes cause friction, in this case we fortunately had the support of the embassy, and we managed to avoid conflicts in our initial meetings and throughout the operation. Midsummer in the European capital, the embassy was very quiet. Personnel were transferring out and vacationing. The ambassador was away, and the deputy chief of mission was about to finish her tour. We checked in with the acting deputy, who was working with us on some delicate last-minute legal concerns that threatened to derail the interviews — namely, how to ensure that the people and information from the interviews would be protected from exposure while still allowing us to make use of the counterintelligence information we obtained. The FBI's Office of General Counsel was reviewing the issue, but Jim Baker was out and his deputy was busy filling in for him at the morning executive meetings at headquarters, slowing down the process. I began to worry about the number of attorneys and their legal opinions entering the discussion, thinking of the old joke that the only thing two attorneys can agree on is that the third one is wrong.

We waited. Morning turned into afternoon. The legal back-and-forth dragged on in transatlantic phone calls and email. As the hours slipped away, I was gripped by a nagging worry that any one of the parties in this delicate inquiry would change his or her mind.

And then, a breakthrough. During an afternoon conference call with our national security attorneys, we got enough of a window to go. *See if you can get the other party to agree to some changes in these areas, but if you can't, we're okay,* the attorneys said. *Good luck.*

Luke and I grabbed our notebooks, and along with my old friend, the ALAT, walked out of the embassy's security perimeter. Outside, we eyed the threatening clouds, praying that the skies wouldn't open up on us. We walked three blocks, then hailed a cab to take us to the agreed-upon interview site. We had the cab drop us a few blocks away and, parting company with our ALAT after safely arriving at our destination, walked to where the interviewees waited. Finally we were able to ask questions.

REDACTED

When the interviews ended after several hours, Luke and I walked toward a major street to look for a taxi. Dusk was approaching, and a sense of deep unease, a heaviness, hung over both of us. The feeling was different from our initial shock when we received the intelligence about Papadopoulos's information. This time it was the weighty burden of confirmation that the report had conveyed the allegation accurately. We had eliminated the possibility that someone had misheard something, and we had satisfied ourselves about the accuracy of the initial reporting. The interviewees were clearheaded, intelligent, and precise. We couldn't get back to Washington quickly enough with the intelligence we had gathered. We returned to the embassy to package up our notes and book seats on the next flight to Dulles.

I would love to weave a compelling story here about the interviews. With whom, and how many. About the atmospherics — where we sat and how we talked. Whether music played in the background. What the body language of the interview subjects was like, and what they wore. Whether we sat by a window as rain drummed on the glass or hunched over a table in a back corner. But there's no way for me to describe the interviews. What was said, how it was said, who was there — none of this information has been released by the government. What's more, the interviews remain one of the undisclosed aspects of our investigation, which requires an unsatisfying void in this narrative.

This is unfortunate, because within that gap lies a story of personal and national bravery against a common enemy, taking place within a unique and storied setting. Without expanding further, I'll say this: the U.S. is blessed with allies who time and again take risks and make profound sacrifices in pursuit of our shared democratic ideals, and the value of their friendship was on plain display that afternoon.

As we made our way back to the embassy, my partner and I had no way of knowing that at almost precisely the same moment 3,500 miles across the Atlantic, two Trump campaign officials, Paul Manafort and Rick Gates, were on their way to the Grand Havana Room, a lounge at the top of a Manhattan building owned at the time by Kushner Properties, to deliver detailed campaign polling data. The recipient was Konstantin Kilimnik, a Ukrainian associate of Manafort's with links to Russian intelligence services, who had just flown into the U.S. from Kiev days after returning from Moscow.

There are few, if any, innocuous reasons for Kilimnik's interest in the polling data — it is far too detailed to serve as a simple display of the strength of the campaign. But it would provide a boon to someone who wanted to know where key voting blocs were, where winning over voters would provide the strongest benefit in the race for Electoral College delegates. Someone like Russian government intelligence officers, who were beginning to place advertisements and targeted posts in U.S. social media. Knowing where to aim their efforts, knowing the issues most likely to appeal to swing voters, would allow them to influence the outcome of our election, shaping it to Russia's advantage.

We wouldn't know for months that the polling data had been shared, but now the clock was ticking even faster than we appreciated.

THE MATRIX

After hurried preparations, Luke and I took a train to the airport to board our return flight. As soon as we disembarked at Dulles, we rushed to FBI headquarters with the results of the interviews, to add them to the ongoing analysis of the initial report.

Central to that analysis, still, was the mystery at the heart of this UNSUB case: who had received the offer of assistance from the Russians? According to statements by both Papadopoulos and Downer, the Australian high commissioner, Papadopoulos had divulged the existence of the offer to the Australians, but who had actually been in touch with the Russians about it? Papadopoulos himself, or someone else, whether on the Trump campaign or otherwise? We wouldn't be able to make meaningful progress on the newly opened Crossfire Hurricane investigation until we could answer that question.

By the following morning, which was Thursday, August 4, the analytic team had made progress on solving this mystery. By making use of our growing matrix — the grid showing all of the data points in the case — we had not only identified a number of individuals who might fit the unknown subject described in the FFG information; we had also compared them against the known evidence about the Papadopoulos allegations to figure out who was the likeliest fit for our UNSUB.

A small team made up of Priestap, Derek, myself, and Mary, who had been the primary FBI attorney on the Midyear team, sat down around the long rectangular table in the conference room across from the Counterintelligence Division front office on the fourth floor of the Hoover Building. Mary was a welcome addition. A precise and tremendously intelligent University of Chicago law graduate, she had worked with us for years. She was assigned to support the Counterintelligence Division, had been a crucial part of the Midyear team, and would prove to be immensely important to the Crossfire group that we were building, the first members of which were now sitting around the table. The huge table could seat 18 but had an odd cutout in the middle, a hollow rectangle into which case files, notebooks, and binders could suddenly topple if one was careless. A bank of monitors, including a secure videoconference system, filled one side of the room. On the opposite side, above U.S. and FBI flags, a row of synchronized digital clocks displayed local times in D.C., Los Angeles, London, Beijing, and of course Moscow.

One of Derek's supervisory intelligence analysts, joined by Luke and a few other members of the team we had already assembled, walked through their preliminary observations, describing potential UNSUBs and their possible Russian contacts as well as these people's histories and their connections to each other, which would be critical to understanding the link — if it existed — from Papadopoulos's statement back to the government of Russia. The analyst, with professorial glasses and a salt-and-pepper crew cut, had a preternatural knowledge of Russian intelligence personnel: their affiliations, history of assignments, predecessors and successors, bosses and subordinates. His scholarly appearance wasn't accidental: he taught college theology classes in his spare time.

I was surprised by the amount of information the analysts had already found. Usually, because initial briefings take place at the very beginning of an investigation, they are short on facts and long on conjecture about all the various avenues we might pursue for information. In this case there were already a lot of facts, and several individuals — not just one — had already cropped up in other cases, in other intelligence collection, in other surveillance activity.

Although I was just hours back from Europe, what I saw was deeply disconcerting. Though we were in the earliest stages of the investigation, our first examination of intelligence had revealed a wide breadth and volume of connections between the Trump campaign and Russia. It was as if we had gone to search for a few rocks only to find ourselves in a field of boulders.

Within a week the team had highlighted several people who stood out as potentially matching the UNSUB who had received the Russian offer of assistance. As we developed information, each person went into the UNSUB matrix, with tick marks next to the matching descriptors.

Papadopoulos was on the matrix, of course. He was not a senior figure in the campaign, nor did he have known links to Russian intelligence officers, but we knew that Trump's foreign policy adviser had been the conduit to the allied nation for the information about Russia. So we couldn't rule out the possibility that Papadopoulos himself was the UNSUB who had first received the overture.

Another of the potential UNSUB matches was Carter Page. Like Papadopoulos, he had been named a member of Trump's foreign policy team suddenly, just before candidate Trump's March 21 interview with the *Washington Post* about his foreign policy plans. Significantly, unlike Papadopoulos, he had a history with certain Russians that the FBI had already investigated. Page's contact with the Russians had drawn the interest of another government agency, which had spoken with him on a few occasions as an "operational contact" (meaning that they could passively debrief him but not task him to conduct any intelligence collection) from 2008 until 2013. I didn't learn of any of that until 2019. Frustratingly and inexplicably, while the Bureau had that information that fall, it was not highlighted in briefings or in FISA applications that would come later.

In my mind, though, the relationship between the other U.S. government agency and Page does not exonerate him, as his contact with Russia continued after his relationship with the other government agency ended. Page's name had come up in 2013 during an unrelated counterintelligence investigation of three Russian intelligence officers the FBI had identified in New York. One of the Russians appeared to be grooming Page to be an asset, although Page apparently wasn't turning out to be a very good one. In fact, FBI agents had secretly recorded the Russian intelligence officer complaining about Page. "I think he is an idiot," the Russian said in the recorded conversation, but that person also said that it was "obvious that he wants to earn lots of money." When the three Russians were eventually charged, Page appeared in the court complaint as "Male-1" interacting with the Russian intelligence officers. Raising more flags for us, Page told us later during questioning that when he read the complaint and realized he was in it, he went to a Russian official, explained that he was the unnamed Male-1, and insisted that he "didn't do anything."

In early July 2016, just a month before my trip to Europe, Page had traveled to Moscow to deliver speeches and presentations, including the commencement address at the New Economic School. His speech raised eyebrows at the FBI, both because the event typically draws much higher-profile speakers than Page and because of the unusual tone of some of his other public comments during the trip. At a lecture in Moscow shortly before his appearance at the New Economic School, he criticized the U.S. for its "often-hypocritical focus on democratization, inequality, corruption and regime change." Answering a question about whether the U.S. was a liberal democratic society, Page told his interlocutor to "read between the lines," continuing, "I tend to agree with you that it's not always as liberal as it may seem," finally concluding, "I'm with you."

Another individual in Trump's orbit who matched many elements of the UNSUB profile was campaign chairman Paul Manafort. He was a much more senior figure in the Trump campaign than either Papadopoulos or Page, and he had a variety of links to Russians, including the oligarch Oleg Deripaska and pro-Russian Ukrainians. Manafort was widely reported to have been brought to the campaign because of his skill at delegate wrangling, which would be crucial if the Republicans ended up in a contested convention. News had emerged two days earlier from the Republican convention in Cleveland, Ohio, that the Trump campaign, which had been relatively quiet on party platform matters, had suddenly begun advocating to strip language from the platform that called for military aid to Ukraine. Manafort's close ties to former Ukrainian president Viktor Yanukovych, who had been driven into exile in a 2014 uprising, and his Moscow-centric Party of Regions, coupled with the campaign's unusual engagement with this relatively obscure party platform item that would benefit Russia, guaranteed that we would watch Manafort closely.

Apart from Papadopoulos, Page, and Manafort, we had identified one other candidate for the UNSUB: former U.S. Army lieutenant general Michael Flynn, a senior adviser to the Trump campaign on national security matters. Flynn had been dismissed as head of the Defense Intelligence Agency. As described by former secretary of state Colin Powell, Flynn had been "abusive with staff, didn't listen, worked against policy, bad management, etc. He has been and was right-wing nutty every [sic] since." We also knew Flynn had engaged in a variety of contacts with the Russian government, including GRU leadership, and had also attended a 2015 dinner in Moscow celebrating the 10th anniversary of the state-sponsored television network, Russia Today, more commonly known by its abbreviation, RT, now known to be a Russian propaganda outlet. Wearing a tuxedo, he sat in the coveted seat immediately to Putin's right.

Things were getting complicated quickly, but our efforts still were guided by the same bright, unwavering lodestar. The question we were trying to answer — the identity of the UNSUB — was not about the Trump campaign; rather, it was about the Russians. Moscow was attempting to manipulate our electoral system by stealing compromising information from one candidate and leveraging it in support of another. Either directly or indirectly, the Russians appeared to have communicated their willingness to assist Trump to at least one member of his campaign. Who had originally received the Russian offer of assistance and relayed it to the campaign? Until we had the answer to that question, we would not be able to confidently answer the question of who else on the campaign knew about the offer of Russian assistance and how the campaign had responded, if it had responded at all. And, most importantly, until we had answered all those questions, we would not have a full picture of what Russia was up to — information that could ultimately help our government decide how to respond to the Russians' attack.

Although Page, Manafort, Papadopoulos, and Flynn were our main candidates for the UNSUB, we didn't look at any of the four as "guilty." Simply put, we weren't targeting anyone. Rather, we were doing the work of FBI counterintelligence agents: investigating a credible allegation of foreign intelligence activity to see where it led. Our goal was to get to the root of what Russia had done, what it was doing, and its impact on national security. It started with Russia, and it was always about Russia.

As the investigation continued into its second week, the briefings moved higher up the FBI ladder. We briefed McCabe on August 10, 2016. At that point we had started looking at second-order associates like Kilimnik, Manafort's business partner (and the pro-Ukrainian contact who, we would later discover, had received the Trump polling data from Manafort and Gates), to explore his intelligence connections to Russia. We also began examining the role of Julian Assange and WikiLeaks in the release of the hacked data stolen by the Russians. McCabe, for his part, had spoken with his counterparts at the CIA and NSA; at the end of the week, I began coordinating with them. Although Comey was aware of the broad contours of the case, McCabe wanted us to schedule a briefing with him to provide a more detailed picture of what we knew and our plan going forward.

Finally, on August 11, we briefed three deputy assistant attorneys general within the DOJ's National Security Division. Getting the three DOJ executives together covered each of their individual national security bases — criminal, intelligence, and cyber — and allowed us to convey how imperative it was to keep the investigations under wraps. We began weekly meetings with them to ensure that we were tightly coordinated and that nothing was falling through the cracks — for example, that a piece of intelligence which might have appeared to be an unrelated cyber matter would be brought to the team's attention if there was any conceivable way it could be related to our investigation.

As the intelligence piled up, it became clear that we would soon have sufficient probable cause to seek a FISA warrant — a warrant for wiretapping foreign intelligence targets, which is reviewed and granted by a special FISA court following rigorous legal standards established in the 1970s — for at least one of our four UNSUB candidates. Indeed, we might already have had enough evidence.

By mid-August we had begun discussions with DOJ about obtaining a FISA on Carter Page, based on the Crossfire allegation coupled with historical information. Given what we knew about Page's past — his alleged desire for money, his disdainful comments in Moscow about liberal democracy, and his Russian connections, including links to several intelligence officers who appeared to have been attempting to groom him as an asset — we had a strong argument that we could establish the probable cause needed to obtain that authority. We didn't have the same sort of probable cause with Papadopoulos, Manafort, or Flynn, but we would continue investigating them even as we contemplated taking the next step with our investigation of Page and eventually applying for a FISA warrant authorizing surveillance of him.

(An aside about FISA is merited here. FISA warrants give investigators permission to unsheathe a tremendously powerful surveillance tool which can also be enormously invasive. As such, these warrants are extremely sensitive, and indeed even talking about the existence of particular FISA warrants, particularly in the context of an individual, makes me extremely uncomfortable. If not for the unprecedented declassification and release of information by the Republican House of Representatives and Trump's White House, I would never — on principle and under the law — do so. But because so much has been disclosed, spun, and deliberately misconstrued about this FISA application, I believe it is important for me to include it in this account.)

At that early stage of Crossfire, the judgment about the evidence to support a FISA warrant for Page was a close call. Highly experienced investigators and attorneys with experience of obtaining thousands of FISAs fell on both sides of the question about whether we could demonstrate sufficient probable cause to obtain search and surveillance warrants on Page. Given that we were just a few weeks into the investigation and had already uncovered everything we had uncovered, however, I felt it reasonable to assume that if we weren't there yet, subsequent investigation would yield enough evidence for us to obtain a FISA later. DOJ agreed, and soon assigned an attorney to work with the investigators and begin familiarizing himself with the case and investigative material that would eventually become the basis for an application for a FISA warrant on Page. For now, we were just planning for the eventuality of accumulating sufficient evidence to establish the probable cause needed for the FISA court to grant us such a warrant. Until the time came for us to take that next step, we kept our noses to the grindstone, following the leads where they led — and periodically sharing our findings with the FBI's leadership.

Comey's first briefing from the team was on August 15. His seventh-floor office, which overlooked Pennsylvania Avenue with a view of the U.S. Capitol to the east, opened directly into a conference room, where nine of us from the operational and legal sides of the Bureau settled around the long conference table and organized our notes and other documents as we waited for Comey to emerge from his office at the end of the room. McCabe was out of the office, so Dave Bowdich, then the associate deputy director, sat in his place, across from me and immediately to the right of the director's seat.

Comey loped in from his office and folded himself into his seat at the head of the table, the usual pad of paper and black felt-tipped pen near his right hand. The Obama White House, he told us, had been discussing how to structure a holistic government approach to Russian interference across executive departments. The difficulty, Comey told us, was how to do that without White House involvement — indeed, the White House personnel had demanded that they not be involved — so as to avoid both the fact and the appearance of partisan motives in the effort.

To its credit, the Obama administration was concerned that any statement it made about Russia's actions might be cast as trying to help Clinton and hurt Trump. But its reticence had unintended effects that weren't limited to the effectiveness of the executive branch's response to the Russian attack. I would watch throughout the fall as the White House's concern began to paralyze the process of deciding what and how to tell the American public about Russia's intrusions into our political process. While understandable, the silence and seeming indecision from the sitting administration hampered our nation's ability to reckon with the threat, much less respond in a speedy and adequate way.

It's important to note that both at the time of our briefing with Comey and before and after it, a wide range of actions were going on across the government regarding Russia, and the FBI was involved with much but not all of it. For instance, investigating Russian election hacking spanned the entire U.S. intelligence community. Planning how to counter that pernicious activity involved intelligence agencies, the nonintelligence components of the Departments of State and Homeland Security, and even Treasury and Commerce, which had economic tools like sanctions, tariffs, and preferred trading status to potentially bring to bear on foreign policy issues. But there were bright lines that couldn't be crossed, some of which were unique to the FBI. Because of the independence of the FBI's law enforcement role, for instance, we didn't discuss our investigations of individual U.S. citizens with the White House — nor did we ask for permission to open those cases.

That firewall between the FBI and the White House occasionally caused misunderstandings between the FBI and the CIA. The CIA collects intelligence to support the president; its work exists to support national security goals driven by the White House. Like FBI agents who didn't always understand how to operate abroad, CIA officers sometimes didn't appreciate the impact of the FBI's domestic law enforcement role and the nuanced ways in which that authority required separation and independence from the White House.

I saw a similar gap on display in August, when a colleague in another government agency working on efforts to understand Russian government and intelligence activity told me, "The White House is running this." That comment may have been sincere or it might have been hyperbole, but for that other agency, it was true. It also explains my response: "Well, maybe for *you* they are." (Notwithstanding the conspiracy theory zealously advanced by Trump and his surrogates, my comment was not an acknowledgment that the Obama White House was running or otherwise controlling our investigations of U.S. citizens. As I've shown, that couldn't have been further from the truth.)

Comey's briefing continued, and Derek and I walked through a condensed version of the same Crossfire overview we had delivered to McCabe the week before. We went into less detail with Comey than we had with McCabe, but included the critical point of the initial intelligence we had received, our views so far on the UNSUB, and the people we had initially identified as potential matches: Page, Manafort, Papadopoulos, and Flynn. We talked about Russians, focusing on the individuals in their government, their intelligence agencies, oligarchs tied to Putin, and all the intermediaries that connected them to each other and to our subjects. We covered the investigative path forward — all the things we planned to do, including the potential for obtaining a FISA on Page at some point in the future, and the work we were doing to identify sources who might have more information about our subjects.

At some point I scrawled in the margin of my notebook's briefing material, "As much info as we can as quickly as we can." The note reflected a directive from Comey, similar to what the military calls the "commander's intent." The sentiment of this note meshed with what we were feeling on the investigative end: as with Midyear Exam, the time pressure on us was extreme, and our worries about the upcoming election had returned tenfold. But the similarity ended there. At the end of the day, Midyear was a mishandling case with little if any impact on national security. In contrast, Crossfire was looking into whether anyone in the Trump campaign was conspiring with the Russians — even up to the unlikely worst-case scenario that Trump was a Manchurian candidate. All of us, not just Comey, acutely felt the profundity of the potential threat to our nation's security.

ACTUARY TABLES

The problem with Comey's imperative — and with the urgency all of us were feeling in mid-August 2016 — was that speed doesn't mesh well with good counterintelligence, which is typically slow and methodical. Nor, for that matter, does it aid secrecy. Hasty investigation is more likely to lead to inadvertent disclosure, and we definitely didn't want word of what we were doing getting out, because it might damage Trump's presidential campaign. We were bound and determined not to affect the election, which was now less than three months away. The extraordinary emphasis on keeping our efforts unknown added a third element to our two goals of getting "as much info as we can as quickly as we can": we needed to keep our investigations as quiet and keep our footprint as small as we possibly could.

In counterintelligence there is an inherent tension between conducting an investigation and protecting sources and methods. There is a tradeoff: the faster and more aggressive the investigation, the greater the chance of exposing the source of information. At worst, exposure can lead to an overseas source's imprisonment or even execution. But there are a host of other, lower-stakes negative outcomes, such as inadvertently revealing identifying information about an email account we might be monitoring, getting a source fired from his or her job, or scaring away potential sources for fear of exposure.

So in Crossfire Hurricane, as in other counterintelligence investigations, we often erred on the side of caution, moving in slow, deliberate steps, every action concealed with plausible explanations. Ideally, in counterintelligence investigations we wanted the purpose of everything we did to be hidden, so as to protect both our investigation and all the secret sources and methods that generated our cases. If our team was trying to get our subject's financial records, we didn't want that person's bank telling him about it. Similarly, we didn't want a source to clumsily start asking a subject very direct and aggressive questions, which might raise his suspicions. If a spy deep in the Chinese government in Beijing identified someone in the U.S. working for China, we didn't want to create a mole hunt in Beijing because we had gone noisily stomping around in Washington looking for their penetration of our government.

This need to avoid exposure and protect sources and methods explains a common complaint about counterintelligence work: its tortoiselike pace. But quiet, clandestine counterintelligence investigations require time — the one thing we didn't have in 2016, as concerns mounted about the Trump campaign's contacts with Russia, and as the presidential election loomed before us.

We frequently debated pace versus aggressiveness in Crossfire. One legitimate view held that caution should be paramount. Our foreign ally had been so deliberate and careful in giving us the information that, according to this line of thinking, we should be equally deliberate and careful in protecting the source, given that exposure could threaten the relationship with the U.S., not to mention the Trump campaign, which, as already noted, was also of paramount concern. Furthermore, this line of argument went, almost every poll showed a decided Clinton edge, providing us more than enough time to proceed cautiously, because any guilty parties weren't likely to end up working in or for the White House.

In other words, if anyone on Trump's campaign was improperly working with the Russians, some of my colleagues argued, we could take our time to figure it out. The advocates for this point of view posited that come November, Clinton was likely to be elected, after which the Trump campaign would disband. At that point, the argument went, we could pursue our various investigations wherever their subjects ended up.

I didn't agree. I argued that our decision-making should never be driven by any expectation or prediction of the election outcome. It simply shouldn't be a consideration. The possibility that someone in the Trump campaign — who might be named to a national security post in a Trump administration — might be working with the Russians to secretly put their finger on the scale to tip the election, I observed, constituted a grave threat to national security whether Trump was elected or not. And the American people, whether they were aware of our efforts or not, were relying on the FBI to figure out whether Americans were working with Russia to undermine U.S. national security in favor of their own personal and political interests.

Shortly after we opened Crossfire, on August 15, I wrote a text message that would come back to haunt me but that neatly sums up my thinking at the time. In the message I wrote, "I want to believe the path you threw out for consideration in Andy [McCabe]'s office — that there's no way he gets elected — but I'm afraid we can't take that risk," continuing, "It's like an insurance policy in the unlikely event you die before 40."

My point was that we were confronting a risk assessment like any other, requiring us to balance the likelihood of a hypothetical event with its potential harm. You don't think you're going to die young, yet the consequences of your accidental death — however remote the chance — are so severe that you take precautions like buying life insurance. Similarly, we didn't expect we would have an administration staffed with people with clandestine connections to Russia. Nevertheless, the chance that people we were investigating could end up in critical national security positions, regardless of how unlikely polling suggested a Trump administration would be, meant that we needed to pursue the investigations quickly. The likelihood seemed low, yet the potential harm was so high as to justify action — in this case, moving ahead with our investigations quickly rather than proceeding slowly.

I felt, and still feel, that an insurance policy was an apt metaphor regarding managing risk in our investigations. My analogy was not a hint at some secret, deep-state plot to keep Trump from being elected; that belief is a fever dream that Trump shares with conservative media and conspiracy theorists. As I noted in later congressional testimony, any of us who were privy to the details of what we were investigating that summer and fall, including information still undisclosed to this day, could easily have damaged Trump's candidacy by leaking that information to Congress or the media, or by taking aggressive investigative steps that would have cast a spotlight on what we were doing. None of us did. We proceeded with deliberate speed, just as Comey had instructed, but we also exercised extreme caution in our investigations. It was a difficult balance to strike, and some of the risk we knew we were incurring manifested itself in the mistakes that DOJ inspector general (IG) Michael Horowitz later found in the Page FISA process. Notwithstanding that valid criticism, and as the IG also found, the investigation of Page was appropriate and within the bounds of the law. And in any event, it was a small part of the massive Crossfire effort. I am proud of how we handled it.

At the time I argued that we just needed to do what we usually would: investigate. That's the FBI's job, and something was clearly wrong here, so let's get to the bottom of it, whatever "it" ends up being. Fears over affecting the election had hovered over Midyear, and I didn't want that to happen with this investigation. The pattern seems to be woven into human nature — much later, in 2019, I saw the same anxieties on display during the post-Mueller debate about impeachment, exhibiting a tacit belief that impeachment wasn't merited because the 2020 elections would address the issue. My answer to all these concerns is the same: maybe, but maybe not. Prognosticate less. Do your job more.

As the days went by, our understanding of the breadth of the campaign's contact with Russia continued to expand. I still didn't appreciate just how unprecedented it was. None of us did. As far as we knew — as far as anyone knew — no campaign had ever been so closely linked to Russia, let alone attempted to coordinate with it. I doubt that anyone on the team would have believed that two and a half years later, three of the four of our initial subjects would have been charged and found guilty of federal felonies, including repeatedly lying to the FBI about their contacts with Russians. As for the idea that Trump would have been shown to have had knowledge of and encouraged any of those contacts, that possibility would have seemed absolutely remote — as would the notion that as president he would be impeached for a similar pattern of behavior toward another Eurasian nation. What we knew in the summer of 2016 was bad, but we had barely scratched the surface.

8

Constantly Awake

IN AUGUST 2016, I had never heard of Christopher Steele. I wouldn't learn about him until mid-September, over a month and a half after the UNSUB investigation began and well after we had started the process of seeking a FISA warrant for Carter Page. In time Steele's intelligence reports would make him a household name and give rise to a cottage industry of conspiracy theories, but during the crucial first weeks of our investigations, he was completely unknown to me.

On September 19, one of the supervisors on the Russia team forwarded several reports that he had received from the FBI agent handling Steele, a former British government employee. Steele had a long history with the FBI. The FBI had worked with him while he was still in British government service. Like our Five Eyes partners in the UK, Canada, Australia, and New Zealand, he shared a similar background in training, methodology, and terminology. We all had different accents, but we all spoke the same language — literally and figuratively. After leaving government, Steele founded a company conducting business intelligence, from which he would occasionally pass the FBI information relevant to criminal or intelligence matters that we were interested in. He wasn't unique in that way. The FBI maintains contact with many people who once worked in the U.S. or foreign governments, then moved to the private sector but still provide useful information. These former national security professionals have a variety of motives in doing so, but the predominant one is a continued patriotic interest in America and its ideals. In Steele's case, it was a deep-seated belief in democratic principles that the British share with us, their cousins across the Atlantic.

Contrary to the popular terminology that would come to define it, Steele's material wasn't a dossier, like a bound compendium of collected intelligence, and we never called it that. Rather, he provided a series of in- dividual intelligence memos that we received incrementally over a period of months, derived from different intelligence sources identified only by code letters. The reports covered a variety of activities and actors relating to the Russian government's interference with our election. They included information about members of the Trump campaign and Trump himself. They contained a great deal of information about Carter Page, whom we were already investigating, and Paul Manafort. As we worked to verify the information, we learned that many of the reports were bouncing all around Washington, with multiple people possessing different reports, many of them believing that they had custody of all the reports when in fact none of them had all the information. In other words, the dissemination of Steele's information was a confusing mess.

The opportunities for confusion and chaos increased as we soon realized that information from the reports was making its way into the hands of the media and elected officials. We heard about the information from Repub- licans and Democrats in both the Senate and the House. On the advice of Senator Lindsey Graham, Senator John McCain provided a copy of some of the reports to Comey. As the memos swirled around Washington and made their way to us, we recognized the danger of circular reporting, and realized that we had to carefully review whether we were seeing information from a new source that confirmed what we already had or a rehashed and recycled version of intelligence we had already seen.

In addition, Steele's information on its own presented issues. The public focuses on the reports' lurid details, especially its suggestion of the exis- tence of compromising material — *kompromat* — of alleged sexual activity between Trump and prostitutes, including a particularly prurient episode at the Moscow Ritz. But that wasn't our focus. What was significant about the material was that it confirmed information — contours of events and players — that we knew to be accurate and consistent with our understanding of the way Putin and the Russians were operating. At the same time, many other pieces of Steele's information — most of them, in fact — were devilishly hard to either prove or refute.

Steele's reports arrived in a format that initially suggested they were almost entirely what the intelligence community calls "raw information" — that is, a recitation of what a source said, as close to verbatim as possible, without any analysis or comment. In the months that followed, as we further reviewed and tested the reporting, we came to understand that his reporting wasn't always raw; it also contained analysis or conclusions based on a combination of information. What's more, the material had initially been gathered for a private client, which Steele identified for us as a research firm called Fusion GPS. Fusion was gathering the information for an opponent of Trump's, but at this early stage Steele didn't have any more details. At the time we didn't know anything more than that, including whether Fusion's client was a private entity, a Republican primary challenger, or a campaign-related entity.

Good tradecraft requires healthy skepticism about information, and we had plenty of questions about what Steele was giving us. He was up front about the goal of Fusion's unknown client: looking for derogatory information about Trump's campaign. From the beginning, therefore, we knew that Steele's reports had been intended for use against a political candidate. That directly affected how we viewed the information, because it fell into a fairly common type of intelligence — common enough that there's a formal tradecraft name for it: "information intended to influence as well as inform." Countless FBI and U.S. intelligence community sources provide information to the FBI when they have a personal stake in how that information is used. Self-interest is an incredible motivator, and intelligence agencies use it to the U.S.'s advantage every single day. But investigators have to be aware of how it might alter what a source is telling them, so that subsequent analysis acknowledges the reliability of the intelligence and accounts for any uncertainty. It's as if each piece of data carries its own statistical error bar.

As we reviewed Steele's information, we realized that we could corroborate a few things immediately and refute a few others. It was impossible, however, to judge the accuracy of the vast majority of the intelligence right off the bat. That's typical for intelligence work. HUMINT — human intelligence, in the lingo of the trade — is frequently imprecise, ephemeral, always evolving, and sometimes completely unreliable. Unlike a point of scientific data, which can be measured, compared, and recreated in experiments, human nature is not easily quantified. People's motivations can change, as can their memories, sometimes without their even being aware of it. Good intelligence analysis applies rigorous and critical evaluation to inconclusive, ambiguous, self-serving, or conflicting information. It's an art, not a science.

We examined the Steele information from at least four perspectives. First, we asked what we knew about Steele, his history, and his motivation for providing the information. Second, we looked at what we could prove or disprove about his reporting, to determine accuracy. Third, we drilled down on figuring out identities, access, motivations, and reliability of subsources of the information. Finally we looked at the nature of the information itself. Did it discuss things that had already appeared in the public record, which might indicate that it was simply a creative spin on a newspaper article? Did it generate productive investigative leads, which would both help our cases and corroborate the information in question? Did it contain details that we could prove were accurate, not publicly known, and harmful to the Russians — qualities that, in other words, made it unlikely to be *disinformation*?

That last possibility is a fundamental counterintelligence concern with any source information, and Steele's material was no different. We considered whether the intelligence, or parts of it, was disinformation that the Russians deliberately placed in Steele's hands with the hope that it would make its way to us or the private client. In short, were Steele's reports part of an elaborate active measure intended to confuse us, occupy resources, or achieve any number of other mischievous goals? The Russians do that well—a hard truth that I'd only come to appreciate fully after Lawrence Martin-Bittman had first tried to impress it upon me back in Boston.

Much later, in December 2019, IG Horowitz released a report explaining in great detail what we did to understand Steele's reporting. As a counterintelligence professional, I believe that portions of that report were ill-advisedly and at best negligently declassified by DOJ. If I were sitting in Moscow working for the SVR or in Beijing for the Ministry of State Security, I would love to have that public report arrive in my in-box. I'm not going to highlight those findings here because I think it would be irresponsible, and in any event, the IG's report is a window into what we did and didn't do. My high-level assessment would be this: we never obtained information to suggest that Steele was lying to us. We learned things about his source network that improved our understanding of aspects of the information. And some of his information did ultimately prove false, although we had no reason to suspect he knew it was inaccurate when he gave it to us. In mid-2017, the end of the time I worked with the group verifying and pursuing the Steele material, my sense of it remained similar to how it had stood at the beginning: some things we could corroborate, a lesser number we could show were inaccurate, and with the vast majority we just couldn't say. We treated it all accordingly.

Aspects of Steele's information were applicable to the Crossfire Hurricane investigation, and not surprisingly we made use of them. For that, the FBI has caught unending flak. To critics of our use of Steele's information, I'd simply say that it isn't ideal to have to work with limited information, especially when it comes from a variety of self-serving motivations—but neither is it unusual. If the information was certain and we'd known everything there was to know, we wouldn't have needed to investigate. Sources are rarely angels; in fact, the best information sometimes comes from devils, wrapped in lies and half-truths. We work with it all the time and know how to separate the wheat from the chaff as a matter of course. Welcome to intelligence.

THE APPLICATION

Steele's reporting added to the information we had about Carter Page and was largely consistent with what we knew about his travel to Russia, the Russian organizations with which he interacted, and the substance of these exchanges. To be sure, as we worked through vetting Steele's information, we determined that several elements of his reporting were inaccurate. But the totality of Steele's material, along with all we had collected by the time we reviewed it, ended debate as to whether we had probable cause for a FISA application.

It's important to remember that we were only at an investigatory stage, trying to prove, *or disprove,* Page's relationship with the Russians; he was one of several UNSUB candidates. We weren't at the point of deciding whether to charge Page with a crime, in which case we would have needed to establish proof beyond a reasonable doubt. We only needed to have probable cause — a reasonable basis to believe — that Page was an agent of a foreign power. We had received some conflicting information about Page, in the Steele material and from other sources, but on the whole his actions and Russia connections gave us more than enough cause for concern.

Between the FBI and DOJ, everybody now agreed that we did have probable cause to apply for a FISA warrant to surveil Page. The drafting accelerated, drawing information from many sources, including Steele's information. His was hardly the only source of information for the application; we had developed information about Page from a variety of sources over years of investigations.

What's more, much of the criticism leveled at us for using Steele's information in the FISA application has been fueled by ignorance about how a FISA application works. There is no recipe by which DOJ drafts an affidavit specifying what percentage of information is derived from what source. Moreover, we took great pains to ensure that the FISA court was informed about Steele's background, in the form of a footnote more than a page long. The description was completely consistent with other FISA pleadings. Even so, we had a vigorous debate about it across many levels of the FBI and DOJ, and the FISA application didn't move forward until everyone was satisfied. Like so much else in the investigation, we would have failed in our duty if we had not investigated as we did.

The IG's later report on the origins of Crossfire found that the FBI investigation of Page was justified, as was our decision to seek a FISA warrant. The report also found that the process was imperfect, pointing to several disappointing errors and to basic procedures that needed to be rethought. But apart from one troubling instance in which an individual appeared to add information to an email used as part of a renewal application for the FISA warrant, the problems that the IG found were unintentional and reflect the increased risk that can occur when a small number of investigators do so much important work in such a compressed time frame. Critically, even if we had addressed every issue identified by the IG, I firmly believe we would still have had sufficient probable cause to obtain the initial FISA warrant on Page. Of course I'm not a judge on the FISC, so I can't speak for them. But decades of experience with hundreds of FISAs convince me that the facts we understood about Page were more than sufficient to initiate the warrant.

Nonetheless, in the months following our initial investigation, the intelligence from Steele spawned a cottage industry of wild conspiracy theories. Deliberate efforts to confuse the narrative fanned those theories, including the work of Russian trolls and, I'm disappointed and sad to say, the president of the United States. Those intentional attempts to obfuscate and sow doubt are still ongoing as of this writing, with influential senators still irresponsibly proclaiming that Page's warrant is evidence of "FISA abuse." They are either willfully ignorant or malevolent. In either case, the politically motivated damage they are doing to the FISA process will harm national security for years to come.

The most prevalent misconception is that the "Steele dossier" launched the entire investigation into Russia's election interference, but as I know and as the IG's Crossfire report conclusively demonstrates, it didn't. Another myth: the investigation into Russian campaign tampering began before the FBI received the Papadopoulos information — incorrect. Another myth: Joseph Mifsud, whom Papadopoulos eventually named as the source of the information about the Russians, was controlled by the FBI — he wasn't, and I have no reason to believe that he was directed by any other member of the intelligence community either. Another myth: Downer was part of a prearranged setup by the FBI to frame Papadopoulos — untrue. Another myth: Sergei Millian, an American citizen of Belarussian descent who contacted Papadopoulos in 2016 about starting a Russian energy business, was directed by the FBI — wrong. Another: a wiretap — or "tapp," in Trump's words — was installed at Trump Tower to spy on the campaign. It wasn't.

I am deeply saddened by the knowledge that because of their improper release, my texts may have inadvertently contributed to the spread of these misconceptions. When DOJ cynically released my private texts to the media, among them the "insurance policy" note that I described in the previous chapter, conspiracy theorists outdid themselves in linking each context-free comment to the Russia investigation. Not just domestic actors, either — Russians calculatedly stoked the disinformation fires. Worse, a disturbing number of these wild theories found their way into presidential tweets, lending the tremendous credibility of that office to malignant falsity and misdirection perpetrated by one of our country's greatest adversaries.

It's worth pausing here to touch briefly on one of those wild conspiracy theories. In December 2015, I texted a question about a foreign counterintelligence operation, which in FBI parlance is known as an OCONUS operation, for "outside the continental U.S." My question was whether an OCONUS "lure" had been approved — an operation to bring an investigative target to a specific location at a specific time, usually to be taken into custody. That text, which long predated and had nothing to do with Crossfire, was lumped into a set of records released to the public and was left partially unredacted and completely unexplained. A fringe "news" website seized on this in early June 2018, claiming that the OCONUS lure discussion proved the FBI had "initiated multiple spies in the Trump campaign in

2015." The following day Trump tweeted about it, claiming (while tossing in "incompetent" and "corrupt" for good measure) that the statement referred to "a counter-intelligence operation into the Trump Campaign dating way back to December 2015. SPYGATE is in full force! Is the Mainstream Media interested yet? Big stuff!" It was in fact wildly false. Simply put, there was no FBI investigation into Russian interaction with members of the Trump campaign before July of 2016, let alone in 2015.

Trump and his enablers also seized on problematic aspects of Steele's reporting when it eventually became public. But in doing so they highlighted a deeper challenge with the "dossier": there were problems with Steele's information that had more to do with the atmospherics around the reports than they did with Steele himself or his intelligence. The main one was the way in which the information seized the public, media, and political imagination and came to define what a clandestine relationship between Trump and Russia must look like. Steele's information became a dispositive test. If one could show all these allegations to be true, then there was an awful, impeachable conspiracy; if not, then No Collusion!

In the public debate, the fog of uncertainty around the Steele material somehow enveloped the entire Crossfire investigation, the majority of which had nothing to do with Steele. The Steele material wasn't the predicate for the FBI's Russia investigation. Steele's information didn't play into the case against Papadopoulos or his conviction. His memos would play no role in the investigation and guilty plea from Flynn. The Steele "dossier" played little role in the crimes charged and proven against Manafort, or in our understanding of Konstantin Kilimnik, or in Manafort's sharing of polling data with him. It didn't address the Trump Tower meeting between the campaign and Natalia Veselnitskaya following up on an offer of dirt about Hillary Clinton. Steele had no role in the allegations that Attorney General Sessions had lied about his contact with the Russians. In short, the information from Steele had very little to do with the bulk of the investigation as we eventually came to know it.

But that idea was largely lost in the debate, I think in part because Trump, his attorneys, and his defenders understood that there was enough uncertainty and inaccuracy in the Steele information that casting doubt on it would allow them to misleadingly claim that there had been no improper relationship with Russia. The media failed in this regard as well: many reputable reporters got caught up in the specifics of the Steele narrative without focusing enough on the problematic aspects of the Trump campaign's connections to Russia that were already known, sitting in plain sight.

Later in the fall of 2016, media reports surfaced about Steele's material having been provided to the FBI. Clearly he had not kept his relationship with the FBI confidential. That's a key component of being a source, and we closed Steele as a result. It's important to note that we did so because he was a control problem, not because his information was bad. He told others about his confidential relationship with us. I'm not sure why he decided to take the actions he did, but there is some internal consistency to it. If he believed even a small portion of his reporting, he probably wanted to do everything in his power to make sure the Russians weren't getting a compromised candidate elected to the presidency of the United States..

HEARTS, MINDS, AND TWITTER

While we pressed forward with our counterintelligence investigations of Trump's associates, our colleagues in the Cyber Division of the FBI watched with alarm as the Russians continued to probe our election infrastructure. They shared updates with us in daily meetings, but unfortunately one crucial prong of the Russian assault fell between our traditional areas of responsibility and caught the entire U.S. intelligence community completely unprepared: Russia's exploitation of social media.

The ability to find and place vast amounts of information on the Internet profoundly changed the art of counterintelligence, particularly in an area of intelligence relevant for the digital age: open-source intelligence, or OSINT. A wide-open category of publicly available information that includes everything from newspapers to professional conferences to think tanks to Facebook posts to 4chan, OSINT is hardly new; intelligence officers and FBI agents have always sought out newspaper articles and specialty journals,

attended speeches and conferences, and listened to radio and TV broadcasts to gain information. Today the difference is the vastness of the amount
of information that is publicly available because of the Internet. In 1994
slightly more than 2,700 websites existed; a year later there were 23,500,
which exploded to about 17 million by 2000. In 2020, well over a billion web
pages populate the Internet.

Social networks saw the same kind of explosive growth. After the founding of Myspace in 2003 and Facebook in 2004, the two companies battled
for supremacy of social networkers. Facebook decisively won that battle
and as of March 2019 claimed 1.56 billion daily and 2.38 billion monthly users, who used the site for social networking, shopping, advertising, political
organizing, and reading the news. Almost 70 percent of American adults
use Facebook, many of them on a daily basis: chatting with friends and
classmates, mentors and strangers, gathering news they think is relevant to
the world, following links to read about happenings around the corner and
across the globe. That number is even higher for YouTube, which 73 percent of all adults in the U.S. view, surging to 91 percent of 18- to 29-year-old
adults. Of course, YouTube is global, with 1.8 billion users (as of this writing), who look up how-to videos, catch snippets from late-night comedy
shows, or watch old sitcoms. Or maybe that's how they watch their favorite
political figures deliver stump speeches or learn about new religious or political ideas that don't appear on the local newsstands. Maybe it's how they
connect with fringe ideas and conspiracy theorists, or dive down the rabbit
hole of political and religious fanaticism.

Adults in the United States use YouTube and Facebook far and away the
most, but there are many others that have claimed significant slices of the
social networking pie. Consider this: the very first tweet was sent on March
21, 2006, at 9:50 p.m., from Twitter cofounder Jack Dorsey: "just setting up
my twttr." By the summer of 2019, Twitter had grown to an estimated 157
million daily users. In the United States, about one in four adults uses Twitter. All of those people may be broadcasting what they had for breakfast or
pictures of their cats, but many of them are also sending out links to articles
and newscasts they agree with, retweets of people they admire, and political

statements that may be trying to advance or enhance their view of their world. Who they follow and who follows them also reflects on who they are. Maybe they comment, maybe they share, maybe they recommend.

One of the reasons that social media has so changed the landscape of counterintelligence is that social networking represents the opposite of coercion. In the past it might have been difficult to convince a reluctant but potential source to turn over a document, to name a public official, to show up at a radical book group that meets in a basement at night. But in the age of the Internet, social media presents a self-selected inviting face of like-minded people, a cozy virtual clubhouse with which one can interact without leaving the house, friends swapping ideas and political philosophies. A tiny bump of serotonin that comes from seeing a tweet or a Facebook post liked hundreds or thousands of times is often incentive enough.

Importantly, from a counterintelligence perspective, social media also makes it more difficult for people to recognize, let alone believe, that they've been duped. In the context of the analog, pre-Internet intelligence world, most people prefer to believe that they're not working with an intelligence officer, even if they have suspicions. Most would rather believe, for example, that they have a friendship with a professor at a foreign university. One of the hardest aspects of counterintelligence work is figuring out whether an asset knows the true identity of the intelligence officer he or she is working with: whether the pretext serves as enough of a subconscious fig leaf to allow the asset to maintain a self-denial about working for a foreign nation, or whether the person is blissfully unaware of the true nature of the officer-asset relationship. By the same token, people today are deeply resistant to the notion that by sharing dubious stories on Facebook, they are furthering the goals of a foreign adversary. Many people, in fact, seem to resent that suggestion, because of the intimation that they're gullible, easily tricked, or consciously aligned politically with that foreign adversary, even if none of those things are true.

The counterintelligence implications of social media were lost on me in the fall of 2016, as they were on most people in my division of the FBI. But we were about to get a rude awakening.

In the early fall, we received word that a private social media expert was conducting research and had created a presentation we needed to see. Three of us got into a bucar and drove out to the northern Virginia suburbs.

Our destination was an anonymous building like any other among the Washington Beltway infrastructure supporting the government's national security apparatus. Nestled into the suburban landscape with neat but sparse landscaping, the newish building was about eight stories of wide glass windows looking out over acres of parking. Its facade bristled with CCTV cameras, a wide security setback lay between the building and vehicles, and a discreet but robust presence of armed guards kept watch. We walked into the quiet lobby, received our security badges, and were politely escorted to a conference room.

We were met by the researcher, who introduced himself, dimmed the lights, and dove into the substance of his presentation. No organization charts, no mission statement, no history of the effort we were about to hear about, all of which customarily consume the first ten to twenty minutes of every federal PowerPoint presentation. He went right to the heart of the matter: that Russia had weaponized Facebook and Twitter to manipulate our country's political discourse. It was clear that Moscow was targeting the Obama administration, but also that its efforts were aimed at exploiting socially divisive issues in our country and using these wedges to deepen and widen existing fissures in the American body politic.

Some of this was already known to the FBI. Throughout the summer we had been aware that Russian-affiliated news organizations like RT, Sputnik, and others had published a variety of pro-Kremlin perspectives about election issues, attacking the thoroughness of the Midyear investigation and broadly questioning the integrity of the presidential election. Those outlets would invariably post their stories on Twitter, and they would be retweeted and incorporated into articles written by more mainstream, or at least not government-of-Russia-affiliated, media organizations.

But the social media expert presented a disturbing twist to this story. His presentation laid out a study of the propagation of the material on Twitter, a study that had discovered a recurring pattern: in some cases, tweets of Russian-affiliated news stories were retweeted by a particular user, then retweeted by a group and retweeted yet again. He visualized this through an evolving graph of the nodes of users, retweeting to others like cancer cells metastasizing. A key component of his analysis was that a defined set of actors was working together in this effort in the same way over time, suggesting that it was a coordinated campaign in pursuit of a specific goal: spreading Russian disinformation.

We left the briefing disconcerted on two levels: first, that Russia was using social media in this way, and second, that we had to learn about it from a private individual. If anyone in the Bureau was tracking such activity, it wasn't making its way into the briefings for senior FBI management, because if it had been, it would have been relayed to us. As our analysts reached out to their counterparts at other intelligence agencies, we were similarly disappointed to learn that ignorance of this behavior permeated the entire U.S. intelligence community. No one was paying attention to it, or if anyone was, it wasn't getting nearly enough attention. The concerted Russian social media effort simply wasn't something that, at the time, was a significant focus of the government. That was about to change.

We had received a crash course in Russian social media manipulation, but all it had done was reveal to us the depth of our ignorance — and the Bureau's. There was no understanding within our team, or among anyone in the FBI or the intelligence community with whom we were working, about the scope of the targeted manipulation Russia was accomplishing using Facebook. The FBI's Cyber Division was aware that the Russians were directly targeting our election infrastructure, but not — as far as we knew — that they were also targeting specific voters directly using social media. To the extent Facebook was aware of what was going on, its managers hadn't shared it with us. Our analysts now were racing to understand the depth, breadth, and impact of Russian activities on social media, but we were playing catch-up.

Equally worrisome was the way Trump used information from Russian-affiliated sites on the campaign trail. We didn't think it was necessarily nefarious; he just amplified evidence to buttress a viewpoint, however wild or incredible, that he wanted to insert into the debate. But it troubled us that he was willing to use what amounted to Russian disinformation in pursuit of those ends. The information wasn't coming from CNN or Fox—it was coming from places like RT and Sputnik, outlets that were clearly closely affiliated with Russia.

Similarly, we knew that WikiLeaks had released materials that the Russian government had stolen from the DNC and the Clinton campaign. By late summer 2016, the public did too, thanks to reports in the press. But Trump and his campaign didn't seem to care; the stolen material was helpful to them, and he mentioned it. A lot. Over the course of 2016, Trump made reference to WikiLeaks over 135 times on the campaign trail. From a counterintelligence perspective, it was problematic that a presidential candidate would use material stolen by a hostile foreign adversary for his own political gain. From a patriotic perspective, I wasn't just worried about a candidate relying on actors outside the U.S. to help his presidential prospects—I was repulsed.

"INSIDIOUS WILES"

FBI personnel weren't the only ones watching what the Russians were doing throughout the summer of 2016. The Obama White House was watching too.

At the beginning of September, national security advisor Susan Rice pulled Comey aside at a White House meeting with a suggestion. It would be nice, she told him, if someone in the government could publicly convey what the Russians were doing. A government voice who was independent, not a partisan, a trustworthy neutral party who could raise the alarm for the American people. Someone like Comey, for example.

Comey brought the request back to FBI headquarters for discussion, musing about the idea of writing something for publication. While we couldn't stop what the Russians were doing, we could share what we were seeing, to throw light onto their actions. Give the public accurate information so that people could make up their own minds about what they were hearing from the candidates and in the news, about what was credible news and what was really foreign propaganda. Inoculate them with the truth.

Comey asked us to take a crack at writing something in his voice that he could publish as an editorial. Three of us started with a classified analysis that Derek and his experts had written and shaped it into an unclassified essay that sounded like Comey. It laid out what Russia was doing, noting that while the tactics were new, Russian attempts at election interference were not, and urging Americans to think carefully about all the news and social media they were consuming.

While I was researching material for the editorial, I ran across a re-markable quote from George Washington that seemed chillingly prophetic, which began, "Against the insidious wiles of foreign influence (I conjure you to believe me, fellow-citizens) the jealousy of a free people ought to be constantly awake, since history and experience prove that foreign influence is one of the most baneful foes of Republican Government."

Washington's words from two centuries earlier perfectly captured the moment we were in, and have since been quoted by others. We used his wisdom to begin Comey's editorial: "In his Farewell Address to the na-tion, President George Washington warned against the 'insidious wiles of foreign influence.' He wrote in response to European efforts to inject false narratives, sow divisiveness, undermine confidence, and weaken U.S. in-ternational standing in political discourse." As we progressed through the few hundred words, we struggled with the conclusion, arguing ideas about sunshine and truth, but we couldn't nail the perfect turn of phrase to pull it all together. We sent what we had to Comey.

The next day he returned a printed copy of our work. With his black felt-tipped pen, he had overwritten our ending with his own, deftly crafting a conclusion that we had not been able to:

> In the same Farewell Address, Washington taught us that "a free peo-ple ought to be constantly awake, since history and experience prove that foreign influence is one of the most baneful foes of Republican Government." The FBI is working closely with our federal, state,

and local colleagues, asking them to double-check defenses, and to continue to apply the disinfectant of sunshine against foreign actors seeking to gain advantage from the shadows.

Comey took the editorial to the DNI, but the White House remained concerned about appearing partisan by discussing Russian interference. The fact was that of the two leading candidates, only one was benefiting from Russia's efforts. Unfortunately, this made the topic of Russia's election interference a third rail for the Obama administration. While we wanted to show the public what Russia was doing—and I sensed that the administration wanted to as well—our desires were offset by the huge counterweight of the need to avoid an appearance of partisanship by the Democrat-led executive branch. Trump played directly to this concern, making claims of a "fixed," illegitimate election—sometimes buttressed by propaganda pieces originating in Russian-controlled media outlets—a central part of his campaign.

Infuriatingly, the Republican Party seemed all too happy to help wall off this issue too. Around the same time in early September that we were drafting the editorial, Comey joined White House personnel and the secretary of homeland security, Jeh Johnson, in briefing congressional leaders about Russia's election interference, seeking support for a bipartisan public statement to raise awareness of what the Russians were doing. The congressional response went beyond the expected party divide, showing the first significant signs of some Republicans' willingness to place their party above country. Senate majority leader Mitch McConnell voiced skepticism about the briefing and refused to issue a joint statement naming the Russians.

After the aborted editorial, the Obama administration remained paralyzed about making any kind of public statement about election interference, fretting that it might be seen as partisan even if it originated from someone seen as neutral, like Comey. Perhaps they assumed the Russians wouldn't be successful. Perhaps they didn't want to tarnish their legacy by appearing to lend support to someone who had been such a bruising primary competitor in an earlier presidential race. Whatever the reason, the government remained silent, and targeted Russian propaganda continued to wash over the American electorate.

Throughout September we were immersed in the Russia investigation — coordinating a meeting in Europe between Steele and agents on the Crossfire team, pushing DOJ to finalize the Carter Page FISA, and identifying investigative opportunities around the world in the cases we had opened on Page, Papadopoulos, Flynn, and Manafort. We hadn't identified the UN-SUB yet, but we felt we were making progress on this question, which was at the heart of the Crossfire investigation. At the same time, innumerable congressional requests for information and documents relating to the Midyear investigation as well as Russian activities against the U.S. piled up.

During the first presidential debate, on September 26, Trump announced, "I don't think anybody knows it was Russia that broke into the DNC." Actually, everyone in the U.S. and Russian intelligence services knew that was the case, and our intelligence community knew it with enough certainty to prompt Obama to personally tell Putin earlier that month to cut it out. The Republican candidate was not privy to this information, but we still had to marvel at how his self-interest trumped his love of country, and constituted a form of Russian disinformation all its own.

Ultimately, the editorial was never published. The administration instead waited another month before finally issuing an anodyne joint DNI/DHS statement on a government website on October 7. It asserted that "the U.S. Intelligence Community (USIC) is confident that the Russian Government directed the recent compromises of e-mails from US persons and institutions, including from US political organizations"; that "these thefts and disclosures are intended to interfere with the US election process"; and that, in the administration's belief, "only Russia's senior-most officials could have authorized these activities." But the statement stopped short of noting how — or in which direction — Russia was attempting to sway the upcoming contest for the presidency. By then the election was exactly one month away.

9

The Laptop

AS CROSSFIRE HURRICANE hurtled forward during the fall of 2016, the day-to-day demands of my job didn't end with Russia work. On top of the UN-SUB case and the other investigations related to Russia's ongoing election interference, I was still overseeing all the thousands of investigations in the FBI's counterintelligence program as well as all the espionage cases handled by the hundreds of investigators in each of the FBI's 56 field offices. Among those cases were the now public investigations of Jerry Lee, a former CIA officer suspected of spying for China, and Monica Witt, a former Department of Defense employee who had defected to, and we suspected was working for, Iran. I was also overseeing the ongoing investigations of Julian Assange and Edward Snowden, as well as countless other highly classified investigations.

Every case was complex and exceedingly delicate. Each one demanded its own attention, time, and precision. At moments the crush of work felt overwhelming. It was an extraordinarily busy time.

And then, in the midst of all of it, a revelation from the FBI's New York field office thrust Midyear back onto my plate as well. On Wednesday, September 28, the field office, known as NYO, noted at the tail end of the weekly videoconference between headquarters and all 56 field offices that an unrelated investigation of former New York congressman Anthony Weiner had, during a search of his laptop, turned up what appeared to be hundreds of thousands of email messages that might be relevant to the closed Midyear investigation. The disgraced former representative, who had resigned over lewd texts and photos and was now under federal investigation for sending more of the same to minors, was married to Huma Abedin, one of Clinton's closest advisers. Suddenly the email investigation we'd thought was closed might prove to be very much open again.

During the Midyear investigation we had scoured the earth for Clinton's email and ended up with multiple copies of most of them, from a variety of different sources. The likelihood that there was something significant that we hadn't already uncovered was slim. Throughout the summer we had received information of a Clinton-related email here or a hacked stash of email there, but none produced anything significantly new. Because of the thorough examination of email that had already occurred, the NYO report didn't seem particularly noteworthy. Still, it met the threshold of a lead that needed follow-up. We hadn't known about Weiner's laptop — nor, for that matter, had Huma Abedin, as she confirmed when we later reinterviewed her. The seventh floor was interested in pursuing it quickly, as was I. That evening I contacted Wayne, the agent who had been the Midyear supervisor, told him about NYO's findings, and asked him to get a team up to New York as soon as possible.

The next day Wayne and Mary, the supervisor and Office of General Counsel lawyer who had worked on Midyear, held a conference call with NYO, after which it was apparent that sending a team to New York would be premature. Wayne briefed me later on the afternoon of September 29 on the substance of the call. NYO still needed to process the digital evidence, and the software kept crashing; nevertheless, they were able to see email that appeared to relate to the investigation. The WFO supervisor told me that NYO didn't have a solid answer yet about what they had. We worried about the accuracy of the initial information we were getting from NYO, because the supervisor had referred to domains we knew did not exist, at least in relation to Midyear, like clinton.com. The NYO supervisor wasn't sloppy or negligent; it was just the reality of the uncertainty of the early stages of computer forensics.

What we didn't know immediately, which would make all the difference, was the appearance of the att.blackberry.net domain, which Clinton had used at the beginning of her State tenure. We had never been able to recover much of the first two months of her email during the time she was secretary of state, some of which may have been among the fabled 30,000 email messages deleted by the Clinton campaign. During that time period, we theorized, she was most likely to have discussed using a private server. If she had, that would provide a critical insight into her state of mind and her intent. Was she advised not to do it? Had someone told her it was allowed?

The problem was that New York's existing search warrant allowed them to look for evidence only of crimes against children — which was the crux of the unrelated case against Weiner — not for potentially classified information related to Midyear. For that we would need our own search warrant, and NYO's preliminary analysis wasn't solid enough to serve as the basis of a sworn affidavit.

Then an honest misunderstanding caused a regrettable delay of the type all too common in large bureaucracies, which has been the cause of much speculation and conspiracy theorizing: the WFO supervisor thought the NYO supervisor would get back in touch when the processing was complete, the NYO supervisor thought Washington would be in touch about getting a warrant, and the OGC attorney thought it was an operational matter for the two supervisors to work out. Everyone thought someone else was working on it. Since digital forensic work can take weeks or even months, I assumed the matter was being handled appropriately and didn't think twice when communications fell silent. I returned to working on the ongoing Russian attack and everything else on my overloaded plate and didn't think about the Weiner laptop for a few weeks.

The pace of the Crossfire investigations quickened. On October 3, we dispatched a team to a European city for the meeting with Steele that we had begun setting up the previous month. I wanted them to clarify his reporting and ask him about other areas of interest, such as whether any of his sources had firsthand information about the topics in his reports. To ascertain the veracity of his intelligence, we needed tangible things: a Russian asset file, a video or audio recording, or access to someone with direct evidence about the reporting. This information was so important to us in part because it would lend further support to our FISA application for surveilling Carter Page. Obtaining a FISA warrant is not a simple process. Although there are exceptions, typically the FBI director must sign off on FISA applications, as does as the attorney general or deputy attorney general. In between them and the case agent lie a variety of agents and attorneys, all of whom play a role in ensuring that the request is appropriate and legal. The application for this FISA warrant in particular received a great deal of attention. In my entire career, only two or three other FISA applications received the same level of scrutiny.

The White House meanwhile revisited the idea of publicly calling out the Russians for the Democratic hacks and presenting the public with some of the intelligence community's key findings as evidence. During an FBI meeting on October 5, Comey expressed concern that it was too late for this kind of public statement; the election was too soon, he worried, and any announcement might be misconstrued as an "October surprise" from the intelligence community — an apparent attempt by the FBI and other agencies to influence the outcome of the election. Moreover, he said, it was by now essentially public knowledge that Russia was behind the hacks, Trump's denials notwithstanding. The administration didn't heed his concern, and two days later DHS and the DNI confirmed that Russia had been behind the DNC hacks in the statement I quoted in the previous chapter.

That date, Friday, October 7, turned into an immense collision of intelligence, politics, and election mischief, all of which triggered an avalanche of news. On that day the *Washington Post* published the 2005 *Access Hollywood* tape in which Trump bragged to Billy Bush about sexually assaulting women ("Grab 'em by the pussy"). Probably not coincidentally, that same day WikiLeaks began posting email stolen from John Podesta, Clinton's campaign chairman.

It was hard to keep up — and if we had known everything else that was going on, it would have been even harder. Just days earlier, Donald Trump Jr. had been swapping direct messages with WikiLeaks, asking in one, "What's behind this Wednesday leak I keep reading about?" WikiLeaks had ended up dumping the information on Friday, of course, not Wednesday, but Don Jr.'s message was evidence that talk of an impending release had been swirling around the Trump campaign. It seemed highly improbable that the WikiLeaks release, coupled with the campaign's foreknowledge of it, had *not* been timed to mitigate the damage of the *Access Hollywood* audiotape.

The campaigns were thundering into the home stretch. At the third and final presidential debate, on October 19, Trump insisted that "our country has no idea" whether Russia was involved in hacking, despite overwhelming evidence from across the U.S. intelligence community. Even as his denials mounted and the election neared, we were getting closer to a final affidavit to support the FISA application for surveilling Carter Page — an investigative step that we hoped would help to clarify whether this presidential candidate happened to have an adviser who was working with the Russians.

The leadership of DOJ's National Security Division, including its chief, John Carlin, had all talked with McCabe about the Bureau's position on the FISA application; the deputy attorney general's office and the adviser to the FISA court had all reviewed it and commented. We had done our part and were just waiting for it to get approved or denied — and, happily, it seemed to be on track for approval. We believed, and I still believe, that there were important reasons to surveil Page, to explore our suspicions that he was working with the Russians.

In the middle of this, a DOJ executive asked McCabe about the Weiner laptop during a meeting on October 25. The question cascaded down the chain within the FBI, and the revived attention made me realize that I hadn't received an update in a few weeks. I scheduled a conference call with the DOJ and the FBI Midyear team the next day. On the call, it quickly became apparent that each party was waiting on the other party to follow up, and that we could — and should — have had this same conversation a week or two earlier.

Everyone was wary of the imminent election, now just under two weeks away. DOJ's nervous counterespionage chief, mindful of the DOJ practice of not investigating political figures around election time, asked whether we should wait until after the election to take action on the laptop. To me and others that would be a mistake, because it would inappropriately take Clinton's candidacy into account in our decision-making, and in any case, the DOJ guidance didn't apply, according to our general counsel, because seeking a search warrant for a laptop that was already in the FBI's possession was an internal investigative activity, which we could conduct without the potential for public disclosure. It was also not until that day that the Midyear team learned from NYO about the backups from the crucial att.blackberry.net domain, that email account used at the beginning of Clinton's tenure that might give us an understanding of her motivations for using a private email server. We all agreed that we needed to get a warrant and that we had probable cause to get one. Now we needed to know what Comey thought. We would find out the following morning.

"SOME PRETTY BIG SURPRISES"

When I sat down to scroll through the news at home that night, I shook my head. Former New York mayor Rudy Giuliani, who was now a Trump adviser and supporter, had told a Fox News reporter that Trump had "a surprise or two that you're going to hear about in the next few days . . . I mean, I'm talking about some pretty big surprises . . . You'll see." Though he was vague, I suspected that he was hinting about the laptop. Giuliani had previously boasted about his FBI sources, although he later denied contact with any active FBI agents. His statement on the heels of our decision to seek a warrant wasn't reassuring.

Giuliani's statement hinted at an undercurrent of politics at the Bureau. The idea advanced by some congressmen that the FBI is some sort of liberal hotbed is laughable to me. The FBI's workforce is largely conservative, and has an institutional mistrust of the Clintons that began in earnest during director Louis Freeh's tenure. I would occasionally catch an earful from agents who knew I was leading Midyear. At lunch with a retired executive one day, I heard, *Pete, you've got to get that bitch.* Stepping toward an open elevator one afternoon, I crossed paths with an agent whom I had worked with long before, when I had first transferred to headquarters from Boston, who said, *We're all counting on you.* My response was always the same: *We're going to follow the facts. If there's something there, we'll find it.*

To be clear, I never felt that anyone was pressuring me to do anything improper. I can't emphasize this FBI cultural norm enough: they wouldn't have done that, and if they had, I wouldn't have tolerated it. I saw this behavior as people with strong personal opinions voicing them. They had their opinions, I had mine (which, unlike them, I kept to myself while talking with superiors and subordinates in my chain of command). We all did our jobs.

This is why Giuliani's alleged FBI sources make me so angry. If people on the inside were talking, that was wrong. They shouldn't have been disclosing investigative information, and they sure shouldn't have been sharing it with a political operative knowing full well it would be used politically to help a candidate.

Similarly, many months later it infuriated me to hear congressman Devin Nunes tell Fox News that "we had whistleblowers that came to us in late September 2016 who talked to us about this laptop sitting up in New York that had additional email on it. The House Intelligence Committee, we had that, but we couldn't do anything with it." The first I heard of Nunes's claim was when he made it, which was in June 2018, and in fairness, I have no idea if that statement is correct. Sometimes members of Congress appeared to have a casual relationship with the truth. But if it was true and someone in the FBI did inform Nunes of the Weiner laptop's email in late September 2016, it's damning evidence of misconduct by whoever did it and should be investigated so that person can be weeded out.

The Clinton email on the laptop wasn't even discovered until September 2016, the same month the "whistleblowers" allegedly approached Congress. The discovery was so new that there was nothing to blow the whistle about. The Bureau was in the early stages of mobilizing to investigate the new lead. Plus, telling Republicans in Congress wasn't blowing the whistle; it was handing confidential investigative information to serve as political ammunition to an opposing party little more than a month before the election.

I know the rumblings about discontent with the Weiner laptop made it to Comey, although I'm certain he didn't get the same level of frank opinion that I heard. But Comey felt it was important enough for him to share concerns about the anti-Clinton sentiment at NYO and resultant concern about leaks with the attorney general. All of us knew the concern was there, and while it wasn't a determinative factor, it played a role in the debate about to come, a debate that would end by upending the final days of the campaign.

As I sat there reading Giuliani's statements on the night of October 25, Trump was delivering real-time denials of any Russia connection at a campaign rally at the Kinston Regional Jetport in North Carolina: "First of all, I don't know Putin, have no business whatsoever with Russia, have nothing to do with Russia. And, you know, they like to say every time WikiLeaks comes out, they say, 'This is a conspiracy between Donald Trump and Russia.' Give me a break." That cascade of denials was wrapped around a lie. Trump's attorney Michael Cohen had been actively pursuing a deal to build a Trump Tower in Moscow. We didn't know that yet, nor did Trump's fans on the tarmac in North Carolina. But Trump knew, and so did the Russians, including Putin. And they undoubtedly took note — more *kompromat*.

SPEAK OR REMAIN SILENT

The meeting with Comey the next morning, on Wednesday, October 26, began on an awkward note. The usual Midyear team, with the addition of Bowdich, gathered in the director's conference room. Derek and I were in our usual seats, while McCabe dialed in from out of the office. When the meeting began, Derek and I had started briefing about the Weiner laptop when Jim Baker cut us off. The previous weekend the *Wall Street Journal* had published an article detailing Democratic campaign donations to McCabe's wife, who had run for the Virginia Senate. Baker suggested that out of an abundance of caution, it might be prudent for McCabe to drop off the call until they could have a discussion about whether McCabe should recuse himself from the Midyear investigation. Comey agreed. It was clear to me from an almost imperceptible pause on the line that McCabe did not.

We all stared intently at the pens and papers in front of us, as if they were suddenly the most interesting things we had ever seen. As quickly as the moment arrived, it passed. With words to Comey to the effect of *Okay, boss, I'll check in with you later,* McCabe dropped off the line, and we resumed the briefing.

There was no disagreement around the table that we needed to get a search warrant for the laptop. To do that, we would need to reopen Midyear. Technically it wasn't closed, because we still needed to wrap up some administrative matters, such as disposing of evidence, but we all knew that the decision to start treating it as an active investigation once again was sure to ruffle some feathers, to say the least. Still, there was no alternative, and there was no serious discussion about waiting to reopen the case until after the election.

The real debate, which would grow in intensity over the next twenty-four hours, was whether to notify Congress, issue a statement, or otherwise comment on reopening the investigation. Comey felt that Congress needed to be informed, since he had told lawmakers on multiple occasions that we had closed the investigation and implied that we would update them if that changed. So he asked me to write the first draft of the notification letter to Congress.

That same afternoon I circulated a draft of a letter and began incorporating edits into it. Essentially it said that an unrelated FBI investigation had uncovered information that we felt was pertinent to Midyear and, as we would in any case, we were going to pursue it. While we didn't yet know whether it would be relevant or how long it would take to determine that, Comey believed that he needed to update Congress based on his prior testimony that the investigation had concluded. The letter would go to Republican chairs and top Democrats in every committee he had briefed.

That evening the final version of the letter to Congress was almost finished. As with Comey's July 5 speech, every word was carefully parsed, every punctuation mark scrutinized.

All of us went home that night with several thoughts. We knew that Congress would immediately leak the letter. We knew that it would help Trump and would be used as a cudgel against Clinton. And we all knew that, despite our efforts to find a third path, there were only two options, both of which were terrible, and both of which were going to throw the FBI headlong into the election yet again, but this time with only a little more than a week until Election Day. Yet it wasn't as if we could withhold the news about the reopened Midyear investigation and risk the appearance of a coverup on Clinton's behalf. Comey's earlier statements and assurances had seemingly locked us onto a path.

In the morning we returned to Comey's conference room, half an hour earlier than the day before. More seats filled around the table with the addition of the heads of our Offices of Congressional and Public Affairs. The mood was quiet; everyone looked tired. There wasn't much to update. I talked through the letter to Congress, but before I could talk about its delivery, Baker suggested that it might be worth discussing opinions about sending the letter at all. The room fell quiet; no one spoke. I couldn't read Baker's reason for raising the issue now, after the decision had already been made. Since no one else spoke up, I broke the silence.

Since I had the floor last, I said, *I'll keep it long enough to raise my concerns about the letter.* There were too many unknowns, I continued. We didn't know what was on the laptop, and based on the investigation we had conducted, it was unlikely that there was anything on it which would change the prior conclusions. We were rushing to a decision about making a statement because waiting would only make matters worse.

My argument felt incomplete, a problem without an answer, a critique without a solution. To my surprise, the deputy general counsel sitting to my left spoke up after I finished. Ordinarily reserved, she laid out her concerns eloquently, including what was unsaid in everyone's mind: *What if the letter ended up helping elect Trump?* After she concluded, Priestap's boss responded from across the table. *I supported sending the letter,* he said, *but now I'm persuaded we shouldn't,* and he then explained his reasons. The discussion expanded into a debate over whether or not we should allow concern about a leak from NYO or anywhere else to influence our decision.

No one was quiet now—everyone around the table seemed to have an opinion. The discussion progressed until everyone had had their say. It became clear that more people were for sending a letter to Congress than were against it. The FBI's silence about Crossfire came up, but not in a contentious way; everyone around the table seemed to understand that we were handling Midyear differently because the two cases were in fact very different. Crossfire was an ongoing, classified counterintelligence investigation that had not been publicly acknowledged; Midyear, by contrast, was a closed criminal investigation about which Comey had testified at length before Congress.

Finally it was Comey's turn. From the head of the table he walked step-by-step through his argument for sending the letter. For him, the issue was starkly binary: speak or remain silent. He began by addressing the concern that the letter might affect the election. He argued that we couldn't allow that to be a consideration. If we did, it would inappropriately and fatally introduce political influence into the investigative decision-making of the FBI. Turning back to the letter itself, he saw both options as terrible choices, but one — not sending it — would be far more damaging than the other. Sending the letter would bring any number of negative outcomes, he acknowledged, but not sending it, effectively remaining silent, amounted to what he considered concealment. That would be unquestionably worse, he concluded.

Though our discussion continued, it was clear that Comey had made his decision. Everyone knew we were about to drop a bomb on both campaigns. A sense of deep trepidation sank over the room.

I have to think that the boss shared our misgivings — but that he also knew that he had no choice. Comey is arguably the most gifted communicator I ever encountered in my 25 years of government service. His natural eloquence went hand in hand with his intellectual agility. Listening to him formulate a position was a bit like watching a craftsman polish a piece of jewelry to perfection: shining it, tweaking it, holding it up to the light, and making improvements until it was complete. In front of that room, he listened to us all and then made one of the most difficult decisions of his career.

Comey and the rest of us may not have made the right decision on that morning, but a dozen thoughtful, experienced people of all backgrounds and political persuasions debated that decision. We challenged every option, and nothing was left unsaid as we tried to do our duty to the Constitution and the American people. By the end, Comey's logic persuaded me, though I didn't feel good about it. If doing the right thing for the right reasons ultimately leads to peace, I'm not there yet.

In retrospect, the decision to send the letter wasn't the hardest choice. The original sin, the fateful decision that placed us on the path to sending the letter on October 28, was Comey's July 5 speech. Once the FBI broke from traditional practice and commented on the investigation, we locked ourselves onto that path. The FBI, needless to say, should never be a factor in any election. If we had been able to look into a crystal ball to see that we were about to inject ourselves into the mix in a way that likely would affect the outcome of the presidential race, would we have made a different decision? That question troubles me to this day, as it has troubled Comey, McCabe, and others on our team, judging by their later comments. I think it's safe to say that most, if not all, of us believed that the letter's publication wouldn't change the election results. Now it seems clear to me that we had more influence than we anticipated, and certainly more than we intended.

In any case, we didn't have time to consider what-ifs at this late stage. Congressional Affairs sent the letter, and as expected, Congress blasted it to the world within minutes. In Grand Rapids, Michigan, for a rally, Trump crowed that "it took guts for Director Comey to make the move that he made in light of the kind of opposition he had where they're trying to protect her from criminal prosecution . . . It took a lot of guts . . . what he did, he brought back his reputation." I didn't watch the speech until later; we were too busy drafting the search warrant and reactivating the Midyear team, dusting off and powering up the custom-built networks that housed the data from the investigation.

After the conference call with NYO, DOJ, and the Midyear team on October 26, DOJ began drafting an affidavit for the search warrant that we knew we would need in order to analyze the Weiner laptop. Working furiously through the weekend, we circulated draft affidavits between the FBI, Main Justice, and the Southern District of New York from sundown Friday until the warrant application was finished on the morning of Sunday, October 30. Application in hand, the Midyear supervisor was preparing to get on an Amtrak train to New York to meet the duty magistrate when we got word from New York that the court would allow him to swear out the affidavit over the phone.

That afternoon NYO turned over cloned copies of the laptop, which our agents drove down to Quantico. We were eager to see what was on it. To a person, we believed that the review would take months.

The computer forensic gurus at Quantico worked around the clock. The hardest part came at the start, waiting for the forensic software to process the laptop, cataloging its contents so that the data could be searched and analyzed. By midweek analysis had begun in earnest, and we were making strides in narrowing down the data for review. We were interested only in email to or from Secretary Clinton. Within that subset, the only relevant email messages were those from the time she was secretary of state. That immediately shrank the pool of relevant email to a much smaller number, in the tens of thousands rather than the hundreds of thousands of messages that didn't include Clinton. The team rotated in shifts at FBI headquarters at the few secure terminals connected to the network we built for Midyear, separating the remaining email into work and nonwork categories, and then, within the work category, into unclassified and potentially classified subcategories.

On Wednesday evening Derek and I received a surprise update: at this pace, we might achieve what we believed to be impossible and complete the review by Election Day. It was a long shot, but it was sufficiently noteworthy to brief up the chain. We began providing daily updates to Comey.

By the morning of Saturday, November 5, we knew we would complete the laptop review before the election. That would mean another letter to Congress, this time describing the results of the review. That also meant discussion about the upheaval that still another letter would cause. But we were already on that path, and there was little debate about sending a letter if we reached that point. We didn't have a choice.

By that afternoon the team had narrowed down the universe of relevant messages to roughly 3,000. We hadn't found anything from the att .blackberry.net domain that might further explain Clinton's decision to use a private email account rather than the State Department system. That left a review of the final 3,000 for anything that was both potentially classified and that we hadn't seen before. Having read everything, the agents and analysts were confident that any new classified information probably wouldn't change our understanding of the case.

I wanted to be certain, though, so Derek, Mary, and I decided to personally re-review all 3,000 to ensure that we hadn't missed anything. Each of us would review a thousand email messages. While the team finished the last stage of their review, Derek and I walked out to buy dinner for the team at a pizzeria a few blocks north of headquarters. As we walked in front of the Hoover Building, we noticed graffiti painted on its facade. It was Guy Fawkes Day, and someone had spray-painted a neon *A* for *anarchy* across the FBI seal. *Everybody's got a complaint,* I thought. We called the FBI police to let them know about the vandalism, then kept walking to get our pizza.

After dinner the three of us hunkered down with several exhausted analysts who stayed behind to help. As we each went through our stack, we quickly realized that there were few messages that might be classified. To save time, we called out the date, time, and subject line over the top of the cubicle walls. An analyst on the other side of the partition would check the email against those already in our possession, all which had received a classification review. Again and again the response came back over the cubicle wall: "We've already got it." Saturday night stretched into Sunday morning. Like grains of sand spilling from the top of an hourglass, the number of unreviewed email messages steadily shrank, from hundreds to sixty to a dozen, until there were none.

It was well after midnight when we double-checked our work, feeling confident in our review and relieved that we were finished. We made plans to go home, rest, and return in the morning. I wrote an email to the Bureau's leadership with the results, and we all packed up our things to go home and try to get a few hours of sleep.

We took the elevator to the second basement level and walked to our cars. As I drove out, I saw movement out of the corner of my eye. My headlights swept across a family of raccoons running to shelter behind snow tractors parked against the southwest corner of the garage.

Hours later we were back at headquarters in a seventh-floor conference room. It was Sunday, November 6, two days before the polls were set to open. Derek and I briefed the executive team with everyone except Comey, and then, after testing our conclusions to ensure they were complete and ready to go to Comey, did it again for the boss minutes later. Finally, just after noon, Jim Baker, chief of staff Jim Rybicki, and I briefed three senior DOJ executives by phone from the general counsel's conference room, walking through our results and the draft of the update Comey planned to send to Congress. After we finished reading it, the principal associate deputy attorney general said, "I think you've threaded the needle." There was a palpable sense of relief.

We hung up and walked back to find Comey in his inner office. On one wall the television aired an NFL game between Comey's beloved New York Giants and the Philadelphia Eagles. Baker recapped the conversation with DOJ, and the head of Congressional Affairs went to print the final letter to Congress for Comey's signature. We were done, our understanding of Midyear unchanged. In less than 48 hours, America would vote for the next president.

THE DECISION

Going into November 8, we knew that one of Putin's overarching goals was to undermine public confidence in the electoral process in a way that would transcend this particular election. The Russian cyber-probing of state election entities and the limited — but still worrisome — number of successful attacks before the election (such as two county election systems revealed by the state of Florida) raised the specter that the Russians might also try to throw doubt on the election results afterward. For example, we were concerned that the Russians might tamper with voter rolls of a close district in a swing state, or simply release a screen shot of a penetrated voter database to call into question the accuracy of poll results. We also knew through classified channels that the Russians had material with the potential to be greatly disruptive, yet they had chosen not to release it. Were they waiting for Election Day? Were they holding it in reserve to discredit Clinton? We didn't know, and we weren't sure what to expect. It remains classified to this day.

In the months leading up to the election, we created a team to monitor news and social media actors who we knew were connected to Russia. But the more we found out, the more we realized how behind we were. Beyond that — and beyond notifying FBI leaders about what our former colleague had told us during his briefing about Russia's social media manipulation — there was very little that our team could do to defend against, much less combat, this threat until we understood more. Even now, with a much greater understanding of Russian manipulation of social media, both the government and private industry are struggling with how to balance First Amendment, business, and national security interests. At the time we had only begun to define what has now become a clearer — if also clearly intractable — problem.

The supercharged political landscape was primed for voting controversy. Just a week before the election, Donald Trump Jr. retweeted a false voter fraud allegation that tens of thousands of ineligible mail-in votes for Clinton had been received in Broward County, Florida. The claim came from the @Ten_GOP Twitter account — the self-proclaimed "Unofficial Twitter account of Tennessee Republicans," but actually a creation of trolls at the Russian-government-linked Internet Research Agency.

But despite all these warning signs, we were still unprepared for what followed. Not only did the Russian election attack in 2016 catch us by surprise, but so too did the impact on the contest that it skewed. In retrospect, we should have seen both shocks coming.

Everyone working the Russian target, both in the FBI and in the U.S.

intelligence community broadly, understood how Russian state power had evolved. We knew about the rise of oligarchs as proxies for the state, the increasing convergence of Russian intelligence services and private contractors, and political manipulation of media by hybrid governmental and private entities. We knew that Putin directed much of the effort through presidential authority, not the Russian intelligence services (a fact that also appeared in Steele's reporting).

The new power paradigm had been on display within Russia, as well as in interference within former Soviet states. But we hadn't internalized it, adjusted our investigations, or updated our collection priorities to gather intelligence that reflected that new reality. We knew about Putin's triad of mischief — undermining faith in our electoral process, helping Trump, and hurting Clinton — but we were ill prepared for its effects.

The structure of the FBI reflected and exacerbated the weakness of the broader U.S. government effort. The people working Russian intelligence and counterintelligence matters weren't the same as those investigating cyber activities or those looking at oligarchs and transnational organized crime. In short, we weren't collaborating as effectively as we should have been.

Reflecting back on 2016 reveals another hard truth: small margins matter in an election in which the total number of Pennsylvania, Wisconsin, and Michigan voters needed to swing the Electoral College would fit in one football stadium. Pundits who argue that it's hard to substantively change public opinion miss the point: when you're dealing with razor-thin margins, it doesn't take much to move the needle. And as much as it pains me to admit it, the Russians weren't the only ones who pushed the needle toward Trump. The Bureau did too.

With the benefit of hindsight, I think it's likely that Comey's July 5 speech and subsequent notifications to Congress on the reopening and closing of the investigation changed the election result. I also think Russia's multi-pronged attack across Democratic databases and social media to help Trump and hurt Clinton changed the result of the election.

Am I certain? Of course not. I've read thoughtful political analyses highlighting other critical factors that may have changed the outcome of an election almost everyone thought would go in Clinton's favor. I don't know to this day which factors mattered and which didn't. It's like arguing which play caused the loss in a close game: the answer is they all did, but none of them alone was determinative. But I do know this: neither Russia nor the FBI should have played a role in determining that outcome. That privilege and responsibility belongs to the American people, and to them alone.

On Election Day I was still at headquarters when the returns began to come in, then eventually watched from home as the final swing states finished tallying their votes. I couldn't believe what I was seeing. No one could —including, it's safe to say, the Russians.

As the night went on, a team of analysts, ironically sitting in a makeshift command post in the Bubble in SIOC, watched as the social media networks of bots and Russian media that had sounded false alarms about election fraud remained strangely quiet, and stayed that way as Trump's upset victory became clear. The Russians must have been as surprised as everyone else. In any event, it no longer makes sense to try to undermine the legitimacy of the election if the candidate you prefer was just elected.

10

Preserve, Protect, and Defend

DURING PHYSICAL TRAINING at Fort Campbell, I hated open-ended runs the most. Our standard routines — eight-lap trots around a track and four-mile loops around the sprawling home of the 101st Airborne Division — weren't exactly fun, but during those runs, at least I could mark my progress as the end point grew steadily closer. I could mentally set mile markers to trick my mind and body into going farther and faster than I felt capable of, knowing how far I had left to go.

No, the worst were the distance runs in which the commanding officer just ordered our battery to set off on an unknown distance down a dirt road through the wild back 40, the installation's sprawling 40-kilometer-long training area. And off we went at a brisk clip toward a truck, far away and out of sight, that would bring us back. Maybe it was three miles away, maybe five, maybe more. Under a blazing Kentucky sun we would run and run, exhausted and dripping with sweat, looking for the truck around every curve, knowing it was parked somewhere up ahead. Lungs burning, legs growing heavy, we never knew where the finish line was or how much energy to conserve to get there until we rounded a final bend and saw the truck parked off in the distance. Sometimes we would gut out the last few hundred yards, aching in anticipation of stopping just ahead.

And then sometimes, suddenly, in the cruelest test of our willpower and endurance, the red brake lights would light up. The exhaust pipe would belch to life and the five-ton olive drab truck would lurch into motion and trundle farther down the road behind a cloud of dust as the finish line disappeared into the distance.

I was reminded of that feeling as I sat in my office at FBI headquarters in the cold dawn of Wednesday, November 9, 2016, the day after Trump's election victory. I was in a foul mood. Like the truck that unpredictably extended a training run, the prospect of bringing our cases to a close had suddenly moved beyond the horizon. At the same time, the apparent national security threat posed by members of the Trump campaign had unexpectedly and exponentially strengthened. Many people we were examining for their illicit connections to Russia were now likely positioned to have access to our nation's deepest secrets and have a direct role in charting our nation's security activity. America's policy toward Crimea, our position in Syria, the posture we would take toward NATO — all areas where the U.S.'s strategic interests were officially at odds with the Kremlin's — were now in the hands of some people we suspected of clandestine relationships with Russia.

It had been difficult enough to investigate and evaluate the counterintelligence risks posed by some in Trump's circle when they were merely running a presidential campaign; soon they would be running an entire country. The whole Crossfire Hurricane team, myself included, was mentally and physically exhausted from the distance we'd already traveled; now we found ourselves staring at the truck's receding taillights, wondering how far we had left to go.

Earlier that morning, rising after a restless night of tossing and turning, I had showered and dressed as on any other day. But everything felt different, sideways and askance. Like the rolling, slow-motion aftershocks of an earthquake, the utterly surreal knowledge that an incoming administration might contain people who were compromised by or colluding with Russia was buffeting me. Half the country was in ecstatic celebration that their iconoclastic candidate had defied all expectations and won. The other half was stunned that this narcissistic, crass television personality would soon lead the free world. My personal values and my privileged professional insights into Trump's campaign put me squarely in this second group.

Not only was I worried about what Trump's election meant for our country, but I was also worried about what it meant for the Russian counterintelligence investigation. The incoming administration would inevitably learn of our investigations before they concluded. Would they honor decades of past practice and let our work continue unmolested? Trump's earlier pronouncements about his desired outcome from the Clinton investigation and his willingness to dispense with the independent rule of law weighed heavily in my belief that he would not.

After I reached my office, I pulled open a desk drawer. From within, an image of Putin in a powder-blue shirt and blue jeans smiled up at me. The glossy photo, bordered with floral designs, adorned the cover of a ridiculous Putin-themed 2017 calendar I had stashed in my desk. I had purchased a stack for my senior team members to try to bring some levity to our overworked crew. Every year someone in Russia issues the propaganda-laden calendar with the president captured in heroic scenes of hyper-masculinity and over-the-top patriotism: Putin bare-chested on horseback! Putin in scuba gear holding an archeological relic! Putin lighting a candle in a church! Putin holding a kitten! It was a gag gift to thank everyone for the sprint we had been running since the end of July. As I stared down into the drawer, a diminutive, malevolent Putin smirked up.

Fucker, I thought, overcoming the urge to slam the drawer shut.

Almost 63 million people had voted for Trump, and nearly 66 million had voted for Clinton. By comparison, only a handful of people knew about our investigation into Russia's efforts to assist the Trump campaign. In the light of day on November 9, that disparity of knowledge took on new significance. The investigation into what had begun as a misfit, long-shot campaign was now an investigation into the incoming administration, which, with the support of a large percentage of the American public, would take control of the executive branch of the United States government in January.

Within weeks the director of national intelligence would contact Trump and his national security team to begin classified briefings. I presumed that the audience for those briefings would include some people on whom we had active investigations. Eventually those subjects might learn that they were under scrutiny by the FBI. This would have devastating consequences for the integrity of our investigations.

On top of that, I couldn't imagine the poisonous politics that would surely follow any public exposure of these cases. Trump certainly wouldn't be pleased when he learned about Crossfire Hurricane — but neither, I knew, would the Republican politicians whose party he now led, let alone the millions upon millions of Americans who had just cast their ballots for him. The backlash was sure to be swift and vicious.

Even armed with that foreboding, I underestimated how ferocious the backlash would be. I wouldn't have to imagine this nightmare for much longer. Soon I would be living it. But for now I saw only one option: run harder.

THE ASSESSMENT

In the days following November 8, the Obama White House ordered the FBI, CIA, and NSA to conduct a comprehensive analysis of Russia's attacks during the 2016 elections. Only a small number of people within the three agencies would compile the study or even know about it, and it would not include any information about the Americans we were investigating. Although it was officially a formal intelligence report known as an Intelligence Community Assessment, in this case bearing the title Assessing Russian Activities and Intentions in Recent U.S. Elections, everyone in the intelligence community shortened that clumsy title to ICA. For the most part, its authors were not people working on our investigations, although I and a handful of people at each of the three agencies both knew about our cases and also worked on drafting or reviewing the ICA.

The necessary secrecy around all these inquiries created significant public confusion about the number and scope of Russia-related FBI investigations going on from the summer of 2016 into early 2017, particularly because some overlapped. The ICA was a stand-alone effort that spanned the entire U.S. intelligence community. Separately the FBI was and is conducting ongoing counterintelligence and cyber investigations to protect against foreign adversaries, and 2016 was no different; we were closely watching what the Russian intelligence services (and those of the Chinese and any number of other nations) were doing in the U.S., including through their intelligence officers here or investigations of that nation's cyber actors. Finally, Crossfire Hurricane and the related individual investigations under it constituted yet another, separate effort.

Another distinguishing factor is that Crossfire Hurricane and its subsidiary investigations were subject to far greater compartmentalization within the FBI and the U.S. intelligence community than most counterintelligence cases. During FBI briefings about Russia, we wouldn't mention Crossfire in the big group, saving discussion about the case until the end, when we could "skinny down," or significantly reduce the number of people participating in that briefing. Limiting the number of people aware of a case, even within the FBI, wasn't unique to Crossfire; highly sensitive investigations originating from a highly placed human asset within a hostile intelligence service, or an exceptionally grave internal espionage investigation like that of former FBI agent Robert Hanssen, have always been handled in a similar manner.

Within the Crossfire team, we discussed whether we should brief the larger group drafting the ICA or refer to our investigations within the report. We decided not to do either one. To me, it was a fairly easy and straightforward decision to keep our investigations separate from the larger ICA group. First, our cases were still active and our traditional practice was not to discuss ongoing investigations outside a small circle of people at the CIA and NSA whose offices are designed to work with us on counterintelligence investigations.

More importantly, expanding the number of people who knew about our investigations would increase the risk of leaks. While no career intelligence officer wants to believe that his or her colleagues might jeopardize an ongoing investigation by sharing information about it with people outside the immediate team — let alone outside the government itself — the fact is that such leaks can and do occur. For the sake of operational security and to protect the rights and reputation of the people who at that point we were merely investigating, we had to minimize the chance of word leaking out.

In short, even though the election was over, we continued to take extraordinary steps to keep our investigations confidential, as far from the public spotlight as possible.

Contrary to cynical counternarratives, we weren't a secret cabal of deep-state agents conspiring against Trump. Just the opposite. The primary purpose of our efforts to keep the knowledge of our investigations limited was to prevent them from becoming public, which, if it had happened, would have done great damage to Trump, his campaign, his future administration, and many people surrounding him. My personal feelings about Trump notwithstanding, I and others made decisions almost daily that were intended to keep the investigations secret and to shield Trump and those in his orbit from the harm that could come with disclosure. Although not optimal from an investigative point of view, as it forced us to move more slowly and cautiously on a case that we all knew was terribly urgent, we felt it was a necessary precaution to protect the new administration and the individuals under investigation. That's one of the many reasons that the argument that I or anyone in the FBI was out to "get" Trump is ludicrous.

Besides the desire to keep the investigations out of the public eye, there was another aspect of the counterintelligence perspective that didn't make it into the ICA. That study was focused only on Russian foreign intelligence, which meant that the analysis would exclude the domestic counterintelligence side of the story. Some at the FBI, including myself, worried that the study's silence about what had occurred in the U.S. would prevent the ICA from providing a holistic view of Russia's attack on the 2016 elections.

I had two concerns about this point in particular. The first was that the ICA would come to be inaccurately viewed as the complete picture of what had happened in 2016. Second, I worried that the comprehensive story of domestic counterintelligence in this period would simply never be told. That fear increasingly preoccupied me as we moved into the new year. It would become even more acute later on, as our counterintelligence investigation into the Trump campaign's and administration's Russia connections expanded beyond what I ever anticipated.

As is customary after a presidential election, President Obama invited Trump to the White House within days of the vote, shortly after Obama's initiation of the ICA. It later emerged that during the meeting, Obama urged Trump to watch out for Flynn. Flynn had served as Obama's chief of the Defense Intelligence Agency but had been fired in 2014, apparently because of mismanagement and temperament issues — problems that Secretary of State Powell had also been concerned about, as mentioned previously. These red flags predated all of what the public would come to know about Flynn's inappropriate relationships and interactions with Russia and his attempts to hide them. But just over a week after the White House meeting, and despite Obama's pointed warning, Trump named Flynn as his national security advisor — a position that does not require Senate confirmation.

While the president-elect was starting to fill out his staff, alarm bells sounded throughout the counterintelligence community about other shady figures who appeared to be working with him in an unofficial capacity. At the same time, within the Counterintelligence Division we changed the structure of how we were managing the cases to reflect the reality that we were going to be working on these cases for longer than we had anticipated. The cases on Page and Manafort, as well as all the work being done with Steele's information, moved over to the Counterintelligence Division's Russia branch and thus out from under my direct supervision. The move was motivated by a recognition of the subjects' continuing connections to Russia. In mid-November, for instance, Trump's former foreign policy adviser Carter Page applied to the transition team. (As it turned out, he would not be rehired.)

The increasingly obvious connections between Page and Russia were apparent even to people in Trump's camp. While the president-elect's transition team took pains to highlight the fact that Page was no longer associated with the incoming administration, doubts were raised and subsequently confirmed by information related to Page's continuing efforts on behalf of the new administration. For instance, on December 8 the Russian-connected Ukrainian businessman Konstantin Kilimnik sent an email intended for Manafort, noting that "Carter Page is in Moscow today, sending messages he is authorized to talk to Russia on behalf of DT on a range of issues of mutual interest, including Ukraine."

Manafort, Trump's former campaign manager, also remained a big concern for us. He had left the Trump campaign in mid-August, but we continued to scrutinize him. And as we did so, a story broke about secret government bookkeeping ledgers discovered in Ukraine after the popular uprising that had unseated Viktor Yanukovych. The ledgers purported to show illicit payments to Manafort, who protested his innocence while jetting around the world pursuing work which, we now saw, might be needed to pay off his alleged oligarch debt. We had known about Manafort's connections to Russia since before the election, and although he was no longer formally associated with Trump, it wasn't clear whether he was unofficially acting as a back channel between Moscow and the incoming administration or simply trying to monetize his prior access to Trump. Either way, he posed an ongoing counterintelligence risk.

As the investigative targets multiplied, Flynn began to recede from our focus, to the point that we had some initial discussion about whether or not to close his case and decided to move to do so. In just a few weeks, however, that discussion would be abruptly curtailed.

The holidays flashed by dimly as the intelligence community put the final touches on the ICA at the end of December. In response to Russia's cyberattacks during the election, the Obama White House prepared punitive sanctions. Over in the Russia branch, my colleagues debated about which Russian intelligence personnel could be expelled from the U.S. as punishment and reviewed Russian properties within the U.S. that could be shuttered. Four days after Christmas, Obama announced the U.S. response to the election meddling: the expulsion of 35 Russian intelligence officials and closure of properties in Maryland and New York that Russian diplomats used for vacationing and other purposes.

Behind the scenes, the FBI, CIA, State Department, and other branches of the government discussed the White House's response and Russia's likely reaction. Expulsions and property closures might appear to make political sense, but in such situations Russia often retaliates with expulsions of its own. So the net effect was less clear — yes, we might get rid of meddlesome Russians who were up to no good in the U.S., but Russia would likely toss out a similar or even greater number of American intelligence officers and diplomats, which would hurt our intelligence posture in Russia, at least in the short term.

So, after poking the bear, we awaited its furious response. But instead, an odd thing happened. Putin did absolutely nothing. Apart from issuing a statement that Russia would "not stoop to the level of irresponsible 'kitchen' diplomacy" (but that the country would "build further steps to restore Russian-American relations based on the policies that President Trump's administration will pursue"), Putin appeared to be following a policy of suspiciously uncharacteristic restraint.

Putin's reaction, or lack thereof, surprised Russia experts in the FBI, the CIA, and elsewhere in the U.S. government. The intelligence community scrambled to explain his nonresponse to the White House.

In the final days of 2016, the FBI developed new intelligence that provided an answer. Without notifying the Obama administration, Flynn — still one of our UNSUB candidates — had spoken with Russia about the sanctions. Weeks before Trump was to be sworn in, Flynn's conversations directly impacted the Obama White House's response to the Russian attack on the election that had just chosen Trump.

Immediately after the Obama administration's imposition of sanctions on Russia, Trump's newly minted national security advisor had secretly spoken with Russian ambassador Kislyak. Flynn asked, on behalf of the incoming administration, that Russia moderate its response to the expulsions and property closures and not engage in what Flynn likened to a "tit-for-tat" escalation.

Flynn's call worked. Russia didn't retaliate. In fact, as Flynn later admitted in the statement of offense accompanying his guilty plea to a series of lies he subsequently told me and my partner, Kislyak followed up with Flynn to explain "that Russia had chosen to moderate its response to those sanctions as a result of Flynn's request."

Our waning attention on Flynn snapped back into focus. His call potentially broke the law — specifically the Logan Act, which Congress passed in 1799 and which forbids any unauthorized person from negotiating with a foreign government having a dispute with the United States. Flynn's call also carried significant consequences, both for him and for the Obama administration, which was in the middle of a tense diplomatic interaction with Russia over its assault on our democracy.

Although we didn't know it at the time, Flynn was worried about the propriety of his actions as well. As he sent texts and email to the campaign about his conversations with Kislyak, he purposely omitted any mention of sanctions. As Flynn later explained to special counsel Mueller's office, "he did not document his discussion of sanctions because it could be perceived as getting in the way of the Obama administration's foreign policy."

One of my DOJ counterparts explained to me that in reality the Logan Act had rarely been invoked as the basis of a prosecution. Legally, there were concerns that the statute was overly broad, but there were also practical reasons not to enforce it at each and every opportunity. As lame-duck administrations prepared to depart after an election, there was a tacit acknowledgment that incoming administrations needed to establish contacts with foreign governments to ease their transition to power. If DOJ didn't exercise prosecutorial discretion, we would end up hamstringing every new presidency and eroding rather than upholding U.S. national security.

But Flynn's case was different; this was not merely a postelection, "looking forward to working with you" grip-and-grin. There was evidence Flynn had conducted a hidden negotiation with a foreign power that had just attacked our elections. While the elected government in power was trying to hold the Russians accountable for their election interference, Flynn was sending a different message: a new sheriff, with new rules, was coming. Perhaps more than anything else, Flynn's secret request and Russia's apparent concession appeared to compromise yet again the incoming administration, potentially providing *kompromat* over Flynn, and possibly even over the president-elect.

At that point we didn't know who apart from Flynn knew something about his interactions with Kislyak. The two men had exchanged numerous texts and phone calls during the end of the year on a variety of topics. We didn't know if anyone had authorized those exchanges. Nor did we know what, if anything, Flynn had told others about his discussions with the Russian ambassador.

Regardless, Flynn now appeared to be in debt to Russia, whose complicity he needed to keep his request a secret. If Trump was connected to Flynn's request — and we didn't know one way or the other — then he might be every bit as vulnerable to Russian coercion as his national security advisor. Like Trump's apparent lies about Trump Tower Moscow, the Flynn-Kislyak call raised the specter of Russia blackmailing the new president of the United States, or at least influencing him via his senior staff — a possibility that could spell disaster for our national security.

Given the gravity of Flynn's actions and the fact that they could explain Putin's surprising lack of response to Obama's sanctions, Comey and McCabe shared an overview of this new information with the director of national intelligence and his senior staff. Once again the finish line moved further into the distance.

While we parsed the revelations about Flynn, we were also preparing Comey for the briefing over the ICA findings that he and the other heads of the intelligence community would be responsible for delivering to President-elect Trump and his incoming White House staff at Trump Tower in New York, and later to congressional leadership in Washington. Comey would be accompanied by director of national intelligence James Clapper, CIA director John Brennan, and NSA director Mike Rogers, who would focus on the areas of the report coming from their respective agencies. The Bureau, for its part, had been busy; the ICA contained a vast amount of counterintelligence information, and we walked Comey through two large tabbed binders full of evidence supporting the assessment's conclusions.

The information from Steele, which had supplemented the FISA warrant application for Carter Page, presented a major dilemma for us as we prepared Comey for the ICA briefing. Steele's material was a powder keg. The ICA's analytic conclusions had pointedly excluded using his information. A large amount of the material could be neither corroborated nor refuted, which had made it difficult to assess, and we had felt that there was little to gain by introducing even those details that we *could* confirm into the otherwise rock-solid analytical conclusions of the ICA. On top of that, the salacious elements related to the president made it even more potentially explosive. Whether these lurid details were true or not, it gave the totality of the reporting a tawdry feeling, and we were concerned that that would detract from other serious allegations that it conveyed.

Within the FBI and among senior leaders in the intelligence community, there was much debate over whether to include the Steele material in the briefing. Although his information had not been used to draw the ICA's analytic conclusions, it had been included in an appendix to the report. Beyond any classified setting, by that point the reporting was also so well known around Washington that it was impossible to ignore.

After extended debate, we decided that Steele's information was too significant not to include in the ICA briefing. We also knew by that point that many people with political agendas were aware of the material, and it was a given that the contents would spill into the public domain at some point. No one wanted Trump thinking that we had the material and had hidden it from him, which might convey the inaccurate perception that we were trying to hold leverage over him.

To emphasize the fact that the ICA had not relied on the Steele material for its conclusions about Russia's intentions, the information was ultimately segregated into a separate addendum at the end of the briefing, to which only a smaller group of people would be privy. The decision was also a gesture of respect—an attempt to avoid embarrassing Trump. With a rueful *Thanks, boss,* Comey told us that Clapper wanted him to deliver the Steele information to Trump in the final, smaller portion of the briefing.

Months later, I reflected with amazement on the bizarre position that this information put us in—an unwinnable dilemma that to the best of my knowledge no other government official has ever had to confront. We had scandalous allegations that we couldn't dismiss or otherwise prove false (which meant, however unlikely, that it could be *kompromat*), coupled with allegations of significant national security importance that we would have been negligent to ignore, all relating to a man whom we now needed to brief about what we thought the Russians had done.

The sheer lowness of it felt unprecedented too. During one prep session, Comey matter-of-factly announced that after some reflection he had decided that he was comfortable telling Trump that the Steele material contained unsubstantiated allegations that Trump had been present while prostitutes at the Moscow Ritz put on a golden showers exhibition, which is slang for sex in which participants urinate on each other. *I cannot believe I just heard an FBI director say "golden showers,"* I thought, troubled. An even more disconcerting thought followed: *I cannot believe the FBI director said that in the context of discussing it with the incoming president of the United States.*

Comey did seem prepared to say it, but it was more than awkward to hear him utter the phrase. Part of it was the incongruity of hearing it from Comey, a disciplined intellectual who wears his strict moral code on his sleeve. But for me there was a different, more profound dissonance to it: the sordid discussion didn't belong in the office of the presidency. The dignity of the office was already being debased, and those words were completely inappropriate and out of sync with the gravitas of the office that Washington, Adams, Jefferson, and Lincoln had once occupied.

While we put the final touches on Comey's presentation two days before the briefing, working out the most effective way to transition between sections, the new president was doing some writing of his own. That day Trump tweeted, "Julian Assange said 'a 14 year old could have hacked Podesta' — why was DNC so careless? Also said Russians did not give him the info!" We had to chuckle when we read his comments on the Hoover Building's ubiquitous wall monitors, many of which were permanently tuned to news channels. Comey was going to face a tough crowd at Trump Tower.

IN WITH THE NEW

On the afternoon of January 6, I took my seat alongside McCabe, Priestap, and several other senior FBI executives around a large table in the
deputy director's conference room, facing a series of video screens. The
room felt cavernous and gloomy, with dim lights shining down on a massive U-shaped table whose dark wood added to the feeling of solemnity.
A screen came to life, revealing Comey in an unguarded moment wolfing
down an unwrapped deli sandwich, looking enormous against the backdrop of a tiny, secure teleconference room. He was still in New York after
the briefing of the president-elect at Trump Tower earlier that day, and he
was scheduled to give a repeat performance later that day in Washington
to senior congressional national security leaders. Known as the Gang of
Eight, the bipartisan group included the Speaker and minority leader of the
House of Representatives, the majority and minority leaders of the Senate,
and the senior majority and minority members of the House and Senate
Intelligence Committees.

The purpose of the video conference was for Comey to back-brief us on
the ICA meeting with Trump. I appreciated his doing it. In a way it was typical of Comey, who was good about informing subordinates about events he
attended. In the case of our Russia investigations, I think he also wanted to
relay his recollections quickly to others while they were still fresh, implicitly
acknowledging that more rather than fewer witnesses might be necessary
in the future.

We sat quietly and listened, asking few questions as Comey described the
reception that he and the ICA got at Trump Tower. Comey noted that the
briefing had unfolded much as he, Clapper, Brennan, and Rogers had practiced the day before. The intelligence community heads had laid out what
the Russians had done and their goals for having done so. Listening to the
briefing were Trump himself; incoming vice president Mike Pence; future
chief of staff Reince Priebus; Flynn; eventual CIA director Mike Pompeo;
incoming national security officials K. T. McFarland and Tom Bossert; and

soon-to-be White House press secretary Sean Spicer. At the end of the main portion of the briefing, Comey said, Clapper had told Trump that the FBI director had something to discuss with the president-elect directly, without the rest of the group. Reince Priebus suggested that he and Pence listen in. Comey agreed that that was fine, but said that it was up to the president-elect. Trump waved off Priebus and Pence, telling them that he would meet with Comey alone. "Just the two of us. Thanks, everybody," he said. Everyone else left the room.

Comey then described how he told Trump that the Bureau had come into possession of the Steele reporting, including the salacious detail about prostitutes in the presidential suite of the Ritz-Carlton in Moscow during Trump's trip to Russia for a Miss Universe Pageant.

At one point Trump interjected a question, asking what year the Ritz-Carlton event was alleged to have occurred. Comey told him it was 2013. As Comey described it to us over the video stream, Trump listened quietly, thinking. And then he said, *There were no prostitutes. There were never any prostitutes.* Trump said he *didn't need to go there,* Comey continued, which Comey took to mean that Trump was famous and powerful enough that he didn't need to pay for sex.

Comey concluded by telling Trump that we knew that a lot of people around D.C., including in Congress and the media, possessed the same reports. Trump wondered why they hadn't been published. Comey responded that he didn't know, but that he wanted Trump to know that the information was out there so he wouldn't be blindsided, as well as to show Trump that he was acting in good faith by letting him know the FBI had also heard the information and we weren't hiding anything from him. Trump thanked him, and that was the end of the one-on-one brief.

As the rest of the group returned to the Trump Tower conference room, Trump tried to make small talk with Comey, telling Comey that he had saved Clinton by not prosecuting her but that the left nonetheless hated him for it. Recounting the comment, Comey didn't seem angry or relieved to have the briefing done. If anything, he seemed as preoccupied and thoughtful as we were, as if the process of retelling the story was simultaneously allowing him to mentally process his impression of what had just occurred.

After Comey finished our back-briefing, he said goodbye and signed off, hustling to make the flight back to D.C. We all sat with our thoughts, looking at the now blank screen. For a moment the room was quiet, everyone trying to absorb what Comey had relayed.

When the conversation finally began, we discussed whether our understanding of Trump's relationship to Russia had changed. It wasn't a planned meeting, but we had most of the FBI's senior leadership in the room — apart from Comey — and it made sense to challenge and test various assessments of what we had just heard. Trump had just been briefed about Russia's attack on the 2016 election, which we had told him was done to help him and hurt Clinton. His reaction — or lack thereof — might shed light on what he did or didn't know. Inevitably, that led to questions about Trump's response to the Steele material. For some in the room, Trump's reaction was inconclusive, because he was vague in his response, neither confirming nor denying it when Comey told him.

I didn't agree. My experience is that people almost universally react with anger and a flat denial if confronted with a false accusation, particularly an outrageously sleazy one. It seemed telling that Trump was nitpicking over whether he ever paid for sex and not whether the episode took place at all in a hotel room in the heart of Russia. Speaking to those around the table, I likened it to accusing a bank robber of a heist involving a shotgun. Instead of denying involvement in the robbery, I said, the thief says, *There was no shotgun. I've never touched a shotgun.*

To this day, I have no idea whether the allegation is true. Later Trump would directly deny the allegation as an innocent person would and ask Comey to investigate it. Still, the fact of the matter is that his immediate, unprepared, and private response, which is often the clearest window into the mindset of an accused person, didn't include a clear denial or disgust. As a counterintelligence professional, I cannot help but be hung up on the many questions that Trump's behavior raised. Why didn't he deny it immediately? If there were never any prostitutes, were there others who might have placed him in a compromising position? Why did the year matter —who knew about that instance, and was Trump trying to figure out the source of the allegation? What might have happened in *other* years?

Personally, I was less worried about whether the president engaged in kinky sex than in the fact that Trump's denial would give the Russians coercive leverage over him if the allegation was in fact true. It would be classic *kompromat*—and not the first, where Trump was concerned. His hidden business dealings about Trump Tower Moscow had already handed the Russians a considerable amount of leverage over him; the moment he lied about the negotiations over the deal to the American people, he had compromised himself. Similarly, when the Russians had granted Flynn's secret request to forestall retaliation for the Obama administration's sanctions, they had gained a possible form of coercive leverage over the new national security advisor, and potentially over his new boss. Now, with Trump about to enter the White House and the list of potential *kompromat* growing longer, what had once been a danger to Trump alone posed a clear and urgent risk to the entire country.

I was reassured that other people in DOJ clearly shared my concerns —and not just about Trump's personal vulnerability to coercion, but also about the disturbing pattern of his and his team's taking actions that would benefit Russia. As we wrapped up in the deputy director's conference room, McCabe asked for a copy of the intelligence we had about Flynn's undisclosed contacts with Kislyak, the Russian ambassador. DOJ had asked about the reporting, McCabe said, and he was going to provide it to a senior official within the National Security Division of DOJ.

I was glad to know that our case on Flynn was getting taken seriously in the upper echelons of the Justice Department. The retired lieutenant general was just weeks away from becoming Trump's national security advisor and had just sat through a briefing of the most highly classified intelligence anywhere in the U.S. intelligence community about the Russian attacks on the 2016 elections. We were not even a week into the new year.

DRIP, DRIP, DRIP

Comey was still making the rounds in D.C., providing private briefings to various required congressional committees about the ICA, when Russian intrigue suddenly seized the public's attention.

On January 10, 2017, BuzzFeed published some of the Steele material online. While it wasn't new to us, it was the Internet equivalent of an atomic explosion. The "dossier," as it was quickly named, contained something for everyone — geopolitical intrigue, sexual deviance, secret meetings, hidden power brokers. The media, regardless of political bent, ran details from Steele's material on a seemingly endless loop, and *Saturday Night Live* rolled out a bare-chested Putin holding a "Pee Pee Tape" VHS cassette at a Trump press conference. Almost immediately, Trump tweeted, "Russia has never tried to use leverage over me. I HAVE NOTHING TO DO WITH RUSSIA — NO DEALS, NO LOANS, NO NOTHING!"

Congress immediately took notice as well. We had heard rumors that several of its members had seen the reporting, including Senator John McCain, who had been concerned enough that he had given a copy to Comey. But the majority of their colleagues on the Hill had not. This was the first time that most members of Congress had heard of it — and they suddenly seemed to be unable to talk about anything else.

The day the Steele material was published, Comey testified before the Senate Intelligence Committee in a long-scheduled appearance. Asked whether the FBI was investigating members of Trump's campaign for connections to Russia, he refused to comment, sparking a storm of furious speculation. Down the hallway on the same day, soon-to-be attorney general Jeff Sessions sat for his confirmation hearings before the Senate Judiciary Committee. Making reference to the material released that day by BuzzFeed, Senator Al Franken asked Sessions what he would do if there was any evidence that someone affiliated with the Trump campaign had communicated with the Russian government in the course of the campaign. Sessions confidently responded, "I have been called a surrogate at a time or two in that campaign and I didn't have — did not have communications with the Russians, and I'm unable to comment on it."

We knew that wasn't true. Sessions had met with Kislyak at a Trump foreign policy speech at the Mayflower Hotel in Washington, D.C., as well as during the Republican National Convention in Cleveland. Perhaps he just forgot, but that seemed extremely unlikely. Besides being the main representative of the Russian government in the United States, Kislyak is a memorably obese man — a Slavic version of the Austin Powers character Fat Bastard. He wasn't the kind of person you'd forget meeting once, let alone twice.

As we knew, Sessions wasn't the only incoming senior Trump administration official who'd had suspicious dealings with the Russian ambassador. The public was about to learn as much too. On January 12, two days after Sessions made his false statement to the Senate Judiciary Committee and BuzzFeed broke the news about Steele, the *Washington Post* published an explosive story of its own. The *Post* reported that Flynn had called Kislyak several times on December 29. Citing a "senior government official" as the source, the *Post* didn't describe the content of the calls, some of which we knew had been about containing the fallout from Obama's punitive response to Russia's election meddling.

The weekend after this story ran, Vice President-elect Pence said on *Face the Nation* that he had spoken to Flynn about the conversations with Kislyak that the *Post* had reported. Pence emphatically asserted that Flynn's conversations "had nothing whatsoever to do with [the Obama administration's] sanctions."

The minute Pence made that comment, I knew it was wrong. That presented a huge problem. Either Flynn had lied to Pence, providing *kompromat* to the Russians, or — far worse — the incoming vice president was party to Flynn's contact with Kislyak and was participating in a coverup by lying to the American people. It felt like an impossibility that the entire chain of command was involved in a clandestine relationship with Russia. But there we were, staring at that possibility.

In the backdrop to all this, the media printed increasingly detailed reports about the contacts between Flynn and Kislyak. I was angry and alarmed about the leaks that fueled this reporting. Angry because it was our classified information that was being leaked, which meant that the sources had to be a person or people within our working group at FBI, or, as I suspected, DOJ, the intelligence community, or Congress. Alarmed because the likely motivations behind the leaks — whether a genuine fear of Trump's team being compromised by the Russians, a partisan desire to discredit a political opponent, or the clock running out on access to internal information for people about to leave office — presented a danger that this drip, drip, drip could become a flood of information about everything we were doing. No one on the Crossfire team, including me, wanted to see the case blown out on the front pages of newspapers across America, at least not until we had finished our investigation and DOJ had decided that our findings merited disclosure to the public through prosecution.

Part of our concern had to do with maintaining the operational security of our investigation, but another part of it stemmed from our values. As law enforcement professionals and civic-minded Americans, we sought to establish and uphold justice and truth. We had seen enough by this point to know that some in Trump's inner circle had an extremely suspicious relationship with the Russian government and that the president-elect himself appeared to be vulnerable to, and very possibly subject to, coercive leverage from it. But if Trump or the members of his administration were not improperly helping Russia at the U.S.'s expense — whether wittingly or unwittingly, whether due to sympathy, blackmail, greed, or anything else — it would be horribly unfair and damaging to the incoming administration if the grist of our investigation was made public. If, on the other hand, they were advancing Russian interests at the expense of America's, that was horrible too, and we had to get to the bottom of it — which would be much easier to do if our investigation wasn't front-page news.

There wasn't much time for us to make more headway on that investigation before Trump took office. Around Washington, barriers and viewing stands were going up along the motorcade route between the White House and the Capitol. The nation's capital was beginning to feel like an isolated coastal town bracing for a hurricane.

As the city prepared for the inauguration, a veritable Who's Who of people with shady Russian connections descended on D.C. as if it were 1930s interwar Vienna. Some of the visits we knew about at the time. Others we didn't find out about until later, but it didn't matter — even if we had known of all of them, there was no way we could have kept tabs on every aspect of what was going on.

Russian tycoon Viktor Vekselberg, who had close ties to Putin and owned a powerful conglomerate called the Renova Group, appeared in Washington and met with Trump's lawyer Michael Cohen. Vekselberg's proximity to Putin caused immediate counterintelligence concern, as did his contact with someone so close to Trump, particularly given Cohen's role as a fixer for the incoming president. Natalia Veselnitskaya, the government-linked Russian attorney who, then unbeknown to us, had appeared at Trump Tower six months earlier to hawk dirt on Hillary Clinton, attended a black-tie inaugural party hosted by the campaign committee of congressman Dana Rohrabacher, who was considered one of the most pro-Russian members of the House. Maria Butina, a gun rights activist from Siberia later convicted of acting as a Russian agent in a covert influence campaign, popped up at inaugural events around town. Konstantin Kilimnik, Manafort's Ukrainian associate with ties to Russian intelligence, arrived and met with the former Trump campaign boss. Sergei Millian, the U.S. citizen of Belorussian descent who by Papadopoulos's account had pitched him on a $30,000-a-month side job if he landed an administration position, attended several inaugural events and dined with Papadopoulos at the Russia House, a downtown D.C. restaurant known for its caviar and vodka.

Even with benefit of hindsight, it's difficult to describe the onslaught of counterintelligence activity triggered by these and other Russian visitors and what it was like to be at the heart of it. I barely slept — Crossfire kept me up late and woke me up early. I worried constantly about what we weren't doing, about what information might be out there but wasn't getting shared, about what leads we had found but might accidentally drop. I skipped workouts and scarfed down horrible food, trying unsuccessfully to force myself to run to the salad bar in the FBI cafeteria instead of grabbing another granola bar from my desk. In the most mundane ways as well as the most profound, the investigation was taking its toll.

It's not that I believed that all of the most awful possibilities were true, first and foremost among them being that Trump was secretly an agent of Russia; in fact, I thought it was likely they weren't. But that didn't change my belief that it was the FBI's job to consider and, where properly predicated, investigate even the worst-case scenarios to protect the American people. That included a sober consideration of whether the man about to be inaugurated was willing to place his or Russia's interests above those of American citizens.

We certainly had evidence that this was the case: that Trump, while gleefully wreaking havoc on America's political institutions and norms, was pulling his punches when it came to our historic adversary, Russia. Given what we knew or had cause to suspect about Trump's compromising behavior in the weeks, months, and years leading up to the election, moreover, it also seemed conceivable, if unlikely, that Moscow had indeed pulled off the most stunning intelligence achievement in human history: secretly controlling the president of the United States — a Manchurian candidate, elected.

That was the context of the meeting we had on that early winter day in 2017, in the small conference room off to the side of Priestap's office. We debated all of the allegations we were aware of and argued about the cases we should open or hold off on opening. It was a fraught exchange. It wasn't just the sheer breadth of the threat — the fact that the list of names written in dry-erase marker on the whiteboard in the conference room filled an entire column. It was the fact that there was reason to put Trump's name at the top of that list.

What had started as a narrow UNSUB investigation that I had assumed would lead us to a single individual within Trump's orbit had instead revealed a broad web of troubling connections between the new president and Russia. And it had opened up a range of possible explanations for this suspicious behavior, some of which we almost couldn't bring ourselves to contemplate. Until recently these had been unthinkable scenarios, beyond the wildest imaginings of most counterintelligence professionals. And these terrible possibilities now included another grim but more immediately testable one: whether the incoming attorney general and vice president were concealing contact with the government of Russia.

"AMERICAN CARNAGE"

There was a stark contrast between Trump's inauguration and Obama's before it. The first swearing-in of Trump's predecessor had been an exuberant, almost ecstatic event, an outpouring of joy and enthusiasm that electrified the city and the nation. By contrast, Trump's was a dark, gloomy event. Speaking on January 20, 2017, to a crowd that barely filled half the Mall, the new president bellowed about "American carnage," lamenting an America rotting from the inside out, rife with violence and despair. Protests in Washington's streets turned violent, with a car set afire and mass arrests. And deception from the new administration began immediately, with press secretary Sean Spicer ordered out in front of the press corps to spin the crowd size, exaggerating the numbers in ways that were easily disprovable. It was clear from the moment Trump took office that his administration was going to be very different from the previous one — and that, contrary to promises that he would become "presidential," the incoming president was not going to be any different from the incendiary candidate America knew from the campaign trail.

For those of us in FBI counterintelligence, the most immediate post-inauguration concern was not the new administration's petty, demonstrably false claim about the crowd size but rather the shadow of a much bigger, ongoing deception — one embodied at that moment by Trump's national security advisor, Michael Flynn. We thought Flynn had conducted unauthorized negotiations with Russia shortly after its attack on our elections — not only that, but that Russia appeared to have granted his request not to escalate tensions with the U.S., a concession that may well have put Flynn, and perhaps his new boss, in debt to the Kremlin. What's more, we knew the new vice president was either wittingly or unwittingly helping Flynn conceal this back-channel diplomacy.

Flynn's lies gave the Russians a form of *kompromat* that they could use against him and perhaps others in the incoming administration. The question was, what were they after? Who else could be in their pocket, and how deeply? What about Pence? What about Sessions? What about Trump?

On the Monday evening following the inauguration, I sat in a conference room with McCabe discussing where we stood with our various investigations and trying to formulate a plan to resolve them. By this time deputy attorney general Sally Yates had heard about Flynn's contact with Kislyak and followed up with a request for more detailed information about their discussions. Between the leadership of DOJ and the FBI, everyone knew we needed to ascertain the motives driving Flynn's behavior, simply because of the extraordinary power he held as national security advisor. The hard question was how to bring the case to a conclusion. To do that we would need to interview Flynn — a prospect that, ironically, was easier to contemplate now than it had been just a couple of weeks prior to Trump's inauguration.

One of the few benefits of the media leaks was that Flynn's contact with Kislyak was now public, and contacting Flynn about it would not disclose anything that wasn't already out in the open. All the attention that contact had received made it logical that the FBI would want to talk to him about the calls. The circumstances would be advantageous in other ways as well; if Flynn's contact with Kislyak had not been public, we would have risked setting him on edge by confronting him about a private conversation that — as far as we knew — he likely would have assumed had been kept secret.

McCabe, several members of the Crossfire team, and I debated how best to interview the new national security advisor. Depending on what you know and what you are trying to find out, there are different ways to approach an interview. One tack is something more akin to a defensive briefing than an interview — essentially warning someone about a threat, with a secondary goal of learning information from them in the process. In a counterintelligence context, the FBI will frequently conduct defensive briefings, particularly of senior government personnel, when it is fairly confident that someone is unwittingly being targeted by a hostile intelligence service.

But in this case a defensive briefing seemed inappropriate, because we knew Flynn might be lying. In the interest of national security, we had to get to the root of his relationship with Russia. Pence had been unequivocal: Flynn had insisted he did not discuss sanctions with Kislyak. While we considered whether Flynn had simply forgotten about discussing sanctions, we all felt that was highly unlikely, given the detailed intelligence we had. And if Flynn was indeed lying, the implications would be far too significant for us to simply give him a warning and walk away with whatever meager intelligence we could collect.

Another option for the interview was to confront Flynn with the information we had about his conversation with Kislyak. But that would require us to lay all our cards on the table, thereby depriving ourselves of the ability to gain insight into Russia's goals and Flynn's state of mind based on what he might do with incomplete information. If he had done something with the Russians we didn't know about, a more circumspect interview might allow us to tease out that information. If instead we shared everything we knew, Flynn would see any gaps in our knowledge, and if he was hiding something else, he might feel that he was safe in continuing to keep it from us. Furthermore, if he had an inappropriate relationship with the Russians, in the worst case, he might convey to them the intelligence information we shared with him.

After a lot of discussion, McCabe set down the ground rules for the conversation with Flynn: no defensive briefing and no sharing the intelligence information. Don't give him any detail about what we know, just interview him about his interactions with Kislyak, which at this point were all over the news. McCabe asked me to do the interview and to take whomever I wanted.

I knew exactly who I needed alongside me when I met with Flynn. It would have to be someone with Russian experience, someone who had the background about the past six months. Leaving the seventh floor, I walked back to my office and picked up the phone to call Luke, the agent who had joined me on my trip to Europe.

Hey, I asked, *want to do an interesting interview with me?*

The next morning, January 24, Luke and I met early to begin outlining the contours of the interview. We soon learned from McCabe that Comey had adjusted our rules — we could quote phrases from the intelligence information to prompt Flynn's memory. To me, it seemed a fair compromise. The nudge could jog an honest recollection or it could extend a fig leaf to allow Flynn to come clean if he had lied. As I wrote in my notes that morning, "Goal: Determine if Flynn will tell the truth about his relationship with the R[ussian]s." The goal was to give Flynn every opportunity to tell the truth about his relationship with the Russians, even if it meant tipping our hand more than we normally would have done.

No date had been set for reaching out to Flynn, but it would have to be soon, since Flynn's desk was now feet away from the president of the United States, privy to the nation's most sensitive of intelligence matters. Still, we figured we would have at least a few more days to prepare.

It turned out to be much sooner than we expected. Around lunchtime that same day, Flynn called McCabe about a separate matter, and McCabe took advantage of the opportunity to set up an interview. McCabe explained that he wanted to send a couple of agents over to talk to him about everything that had been in the news. It was up to Flynn, McCabe continued, whether to have others present: *If you want to call the White House Counsel's office, that's fine, but then I'll need to get DOJ involved.* Flynn quickly said that wouldn't be necessary and unexpectedly proposed to meet with us in the West Wing that afternoon.

Luke and I sped up our interview prep. We had hours, not days.

11

The Situation Room

THE ONLY WAY I've entered the West Wing is through the Situation Room, the part of the White House where the president's most sensitive discussions and operational activities take place. Despite its name, the Situation Room isn't a single room; rather, it's a complex under the Oval Office and the West Wing of the White House. It has its own outdoor entrance, from a gated access road separating the White House from the Eisenhower Executive Office Building, an imposing building that houses the National Security Council staff and various support offices of the president and vice president. Visitors with seniority or who access the Situation Room frequently, such as cabinet members and intelligence agency directors, can drive — or, more precisely, be driven — right up to the sheltered entrance. Lesser visitors, such as my partner Luke and me, have to navigate a ring of lines and checkpoints from 17th Street into the inner sanctum.

That's how we entered the West Wing on January 24. Because it wasn't particularly cold, we considered walking from the Hoover Building, which is about twenty minutes from the White House by foot. But we knew that the White House visitor access software had been acting up, so we decided to take a cab from FBI headquarters to leave ourselves extra time in case of security delays.

The taxi dropped us off at New York Avenue and 17th Street NW, a few blocks from DAR Constitution Hall, with the Washington Monument rising over the Mall to the southeast. At the first security gate we showed our

FBI credentials to the Secret Service officer, who verified that we were on the access list and handed back our credentials along with our temporary White House access badges. Passing through the gate, we walked eastward on State Place toward the White House, past the inevitable line of parked black government SUVs, then turned left to another guard shack. After we passed through the magnetometers of the final checkpoint, we made our way up West Executive Avenue between the White House and the Eisenhower Building to the unremarkable white awning on the right bearing a monotone gray seal of the president of the United States, marking the entrance to the Situation Room.

One of the most surprising things about the West Wing is how compact everything is, and the Situation Room is no exception. Luke and I descended a few steps into a windowless entrance area with low ceilings and a hushed ambiance that felt too cramped to contain the enormous issues that had been discussed within its walls. We passed a Secret Service officer seated just inside the entryway. Having arrived 20 minutes early for the 2 p.m. interview, we sat on small sofas across from the men's restrooms. To our left, a few stairs led down to a canteen and a small dining area, the constantly staffed Watch Office (which monitors events around the world and can put the president in contact with anyone at any time), and several conference rooms, including the one made famous by photographs of President Obama watching the raid on Osama bin Laden's compound.

I had been to the Situation Room several times before. On the best days, I had gone there to provide updates on significant counterintelligence investigations. On the worst ones, it was to deal with grave security lapses, diplomatic imbroglios, or tragedies, to explain what we did and didn't know about the troubles at hand. In either scenario, the briefings were unsparing, detail-oriented, and delivered to officials with a variety of obtuse and extended titles.

But this visit, like so many other experiences I seemed to be having these days, was a first. I wasn't coming to offer information to the nation's top officials. For the first time, I was coming to investigate one of them.

With a high-stakes interview like the one I was about to conduct with

Flynn, it's crucial to have a partner like Luke. Beyond his Russian expertise, Luke was unflappable, the sort of agent who could adjust without missing a beat if an interview took an unpredictable turn. If evidence of alien life suddenly arose during an interview, Luke would seamlessly pivot to asking whether the aliens were green or gray, and how tall. Experienced agents like him can tune out outside distractions and focus their attention with an intensity bordering on obsessive-compulsive. They are also meticulous planners: if they have time to mentally walk through an interview outline six times, then they'll do it a seventh time, in constant pursuit of Vince Lombardi's idea of "perfect practice." We hadn't had time to do our own run-through of the interview with Flynn, but as we sat on those sofas in the bowels of the White House, I could be sure that Luke was mentally walking through key questions and showstoppers that might derail the interview.

As we reviewed our thoughts and our physical notes, I saw Flynn wheel around the corner beside the Secret Service officer. He spotted us almost as soon as we spotted him, and we stood up to greet him. He was expecting us, of course, but in dress and haircut and demeanor — not to mention race and gender, unfortunately — most FBI agents look alike, which probably would have tipped him off to our identities even if we'd been arriving unannounced. He asked if we wanted to start a little early and we agreed.

Some in the FBI held Flynn in high regard, in part for his collegial relationship with the Bureau in the wartime theaters of Afghanistan and elsewhere in the Middle East. And I had to admit that, whatever his political acumen, his personal appearance and mannerisms gave me an almost odd sense of comfortable familiarity — a feeling I often got around police officers and military personnel, but not often with White House staffers. A compact, fit man, he had a direct intensity.

We followed Flynn back past the Secret Service officer, bearing right past an elevator and toward a staircase leading up to the West Wing. In one of my early assignments to FBI HQ, a wise FBI agent detailed to the White House taught me that the best way to judge the importance of any White House staff member was by the proximity of his or her office to the Oval

Office. Now, as we approached the staircase, Flynn asked if we had been to the White House before; Luke responded that he had not, while I explained that my official visits had been limited to the Situation Room. "Let me give you a quick tour, then," he offered.

We went up a flight. We walked across a lobby exiting onto the north side of the White House and turned to the right as we reached the Cabinet Room, the largest conference room in the West Wing and a frequent scene of press conferences where the president is flanked by his cabinet members.

As we walked between the Roosevelt Room and the Oval Office, the doors suddenly opened and two workers, each carrying oil paintings, emerged with Trump in tow, directing them. The president was personally overseeing the Oval Office redecoration, bringing out artwork that Obama had hung and replacing it with his own selections to overlook the Resolute desk — one of which was the portrait of Andrew Jackson that later drew so much attention.

Trump breezed past Flynn, Luke, and myself as if we weren't there, forcing Luke to step to the side as the fine-art caravan blew through our midst. It was a startling moment for me. It was the first time I had seen this president in person, and I was struck that Trump was taller than I anticipated. I was also surprised at how directly he was supervising the workers. He didn't say anything to Flynn, and I don't think he noticed him. Flynn leaned toward me. *You have to remember, he owns a bunch of hotels,* he explained. *He's got a great eye for decorating.* The remark was tinged with admiration and a hint of awe.

As we continued along a small loop on the way to Flynn's office, the easy conversation continued, delving into Trump's lodging preferences on the campaign trail. *We always stayed at Trump properties if any were nearby, but if not, he thought Holiday Inns offered the best quality-to-price ratio,* Flynn offered.

Flynn's large office in the northwest corner of the West Wing was still in a half-unpacked state, with unopened boxes stacked around his desk, others open, spilling out books and files. Uneven stacks of papers sloshed across tabletops, and the walls were still unadorned. Luke and I sat next to each other while Flynn took a seat facing us. We settled in and began the interview.

I asked most of the questions while Luke took notes. To the extent possible, we tried to keep up the easy banter we'd struck up with Flynn on the way into his office. The idea behind any good interview is that it should feel like an informal but purposeful chat — natural and direct, but not awkward or disjointed. As with all well-planned interviews, we knew the critical questions we needed to ask Flynn, but we tried to couch them in conversational terms and tones, as if we were just talking over a beer.

Flynn appeared cooperative throughout the interview. But as it wore on, something strange began to happen. When we asked Flynn key questions about his conversations with Kislyak, questions we knew the answers to — and that *he* knew we knew the answers to — he repeatedly and inexplicably lied. He had to have read the various newspaper reports leaking details about his conversations with Kislyak. McCabe had explicitly told him we were coming to talk about those very conversations, and Flynn had even told McCabe during their phone conversation that we already "probably knew what was said."

Even more puzzling, even as Flynn baldly lied to us, he didn't exhibit any of the "tells" of people who are lying. At Quantico new agent trainees are taught to look for "nonverbal indicia of deception." Pausing before answering. Repeating a question. Crossing arms or legs. Licking lips. Subconsciously bringing your hand up to cover your mouth. Flynn didn't do any of those things. He didn't avoid questions or weasel his way through answers. He just gave us answers that didn't square with what we knew had happened — and what *he* knew we knew had happened.

What's more, Flynn's most unsatisfying answers were conspicuously redundant — and from my point of view pretty inconsistent. He did admit to receiving a text message from Kislyak while on vacation in the Dominican Republic in December 2016 and calling Kislyak the next day. But when I asked Flynn if at any point during the transition period he had talked to Kislyak about the expulsion of Russian diplomats, the closing of Russian properties in response to Russian election-related hacking, and the need for

Russia not to engage in a tit-for-tat retaliation, Flynn responded, "Not really. I don't remember." He said he had not asked Kislyak for Russia to vote a particular way in an upcoming UN resolution about Israeli settlements in Palestinian territories, when we knew that Flynn had done just that. And he said that Kislyak hadn't followed up with any feedback about that request, when we knew he had.

We provided information to try to "help" Flynn remember the things he claimed to be forgetting — particular turns of phrase that he used when talking with Kislyak. It changed nothing. Try as we did — repeatedly — we couldn't get him to budge from the story he told us. He also omitted several things that we learned later, such as repeatedly discussing the calls with his incoming deputy, K. T. McFarland, who was with Trump at Mar-a-Lago when Flynn was in the Dominican Republic.

This is not to say that everything out of Flynn's mouth was a lie. He confirmed that in 2013 he had met Igor Sergun, the head of Russian military intelligence, the GRU, during a leadership program in Russia. He volunteered that he had had a closed-door meeting with Ambassador Kislyak right after the election. When we asked him about a possible conversation he had had with Kislyak in December about the UN vote, he replied, "Yes, good reminder," and even thanked us for reminding him of the call, but then he omitted specific requests he had made of Kislyak and Russia. He framed the call in the context of many others he had made to representatives of Israel, the UK, Senegal, and Egypt to find out how they would vote on the resolution, a context we knew was false.

Luke and I didn't look at each other as the interview went on. As I completed the list of items we wanted to discuss, I asked Luke if he had any questions. Luke returned to a few areas to ask for clarifying information, and then I concluded the interview with a ubiquitous closing: *Is there anything we didn't ask that you think is important for us to know?* There was not, Flynn said.

After we finished the interview, the discussion turned to post-interview small talk about Flynn's impression of the first days on the job. The conversation was as free-flowing as ever, and Flynn even seemed to be enjoying it; indeed, when a staff member stuck his head in the door to tell Flynn his next appointment was coming up, Flynn asked him to push it back. As Luke and I listened, Flynn chatted about the transition to his new position, remarking with what seemed to be wide-eyed sincerity that he had been surprised by the intensity of the political attacks directed at him since the election. *How can that surprise you?* I thought. *You've spent the last several months leading chants of "Lock her up!" at campaign rallies.* I got the sense he simply didn't understand the political context of the world in which he was now operating.

The meeting dragged on. Flynn didn't act like someone encumbered with time constraints or obligations; instead he seemed excited to talk. Turning to a large map of the globe he had been provided during a classified briefing, he began describing some of the intelligence networks depicted on the map. It was a very strange moment. On the one hand, I saw that if he continued in his role, the new national security advisor would be a strong advocate for the FBI's counterintelligence work. But on the other hand, his impulsive eagerness to talk about a classified matter playing out overseas with two FBI agents who happened to be in his office suggested that a deliberate and coordinated national security policy might be an early casualty of the administration.

As we continued to listen to Flynn, it dawned on me that this discussion wasn't wrapping up. It wasn't that it was a long diversion — it was just that individual minutes of the national security advisor's time are precious. Or at least they should be. I became increasingly uneasy with the amount of time Flynn was spending with us, like a taxi driver turning around to face you to make an extended point while his car hurtles down a busy street. *You're two weeks into your job as national security advisor,* I thought. *There have to be a thousand other more pressing things for you to be doing. Please, let's wrap this up.*

I abruptly stopped the conversation and told him we looked forward to working with him and the new administration. *We know you're extraordinarily busy,* I said, *and we appreciate your time.* We stood up, shook hands, and left.

Outside, the late afternoon held the lingering warmth of the sun. Luke and I had been in the White House for about an hour and a half. We strolled slowly down West Executive Avenue, the White House looming on our left. Neither Luke nor I said anything for a few minutes, mulling over what had just happened. I felt a weight upon me similar to what I had felt in Europe after conducting our interviews — a dark sense of foreboding. But unlike in the European capital city the previous August, this time we were walking among monuments to American democracy, institutions set up as a bulwark against foreign adversaries who wished ill upon our country. We had just exited the offices of the most powerful government on earth. I had hoped to leave with a sense of relief that we had been wrong about Flynn. Instead I felt an even greater disquiet about the president's national security advisor.

You up for walking back? I asked Luke.

Yeah, he replied.

We headed south, still lost in our thoughts, and continued along the Ellipse between the White House and the Washington Monument, toward FBI headquarters and the hundred questions we knew would be waiting for us there.

Breaking the silence, I asked, *So, I don't want to prejudice your response by telling you what I thought. What did you think?*

Luke responded with exactly what I was thinking: Flynn hadn't behaved as if he were lying. But what he had said wasn't true.

Between Luke and me, we had conducted over a thousand interviews. Our intuitions were well honed. The fact that we had arrived independently at the same conclusion was telling and reassuring — but it was also confounding. Flynn had known why we were there, and he had to have known what the truth was. What's more, he had agreed to meet with us only a few short hours after we'd requested an interview and hadn't wanted an attorney present for the meeting. He didn't even appear to have told anyone else in the White House that we were coming over. And yet he had still lied — as he later admitted and swore to judges, in court filings, and in statements to other investigators.

Sitting in McCabe's office later that day, Luke and I briefed McCabe, Baker, Priestap, and other senior members of the team. They were understandably doubtful. Like us, they struggled to make sense of how Flynn could have been so forthcoming with statements that contradicted widely reported facts we already knew and yet appear not to be lying.

After we had covered the crucial details, McCabe and Baker left to update Comey. Luke and I stayed behind to keep talking with other members of the leadership team from the Counterintelligence Division and General Counsel's Office. They continued peppering us with skeptical questions, to the point that I began to get slightly irritated.

I understand it doesn't make sense, I explained for what felt like the millionth time. *I understand that his statements are not true. I understand that he probably lied to us. But if he did, he's really, really good at lying, because our careful, experienced observation of him didn't show any of the behavior we would have expected. Or there might be something more complicated in play, some sort of denial or disorder.*

And we're approaching the point at which it begins to feel like you're questioning our competence, I thought to myself.

Flynn's behavior and the logic and motivation behind it were a mystery — one that we didn't solve during the meeting and as far as I know still hasn't been answered. I have to wonder if Flynn was so deep in denial that he'd somehow persuaded himself that certain things hadn't happened. Maybe he forgot, which seems impossible given the avalanche of media coverage at the time and the fact that McCabe had told him exactly what we wanted to talk to him about. Or, if he understood what he'd done, perhaps he was just too naive to know how much trouble he was in, or was so self-assured that he believed that he could wriggle out of it — that he was so good a liar that he could somehow pull it off. It was, and is, a puzzle. Through his early career, at least, he was a respected officer who had made the threat of radical Islamic terrorism his call sign, but this sort of political maneuvering around truth and laws seemed to be beyond him. Certainly he underestimated the sophistication of the nation's intelligence community. He was about to learn about it the hard way.

"I'LL TAKE THAT BACK TO THE DIRECTOR"

The day after Flynn's interview, January 25, I walked across Pennsylvania Avenue with Jim Baker, Bill Priestap, and Mary. We were headed from the Hoover Building to a conference room within the National Security Division of DOJ headquarters, where we were scheduled to brief the leadership of that division and a member of the deputy attorney general's office who worked on national security matters.

Word of Flynn's interview had made its way across the street, and DOJ officials wanted to be kept abreast of all the rapidly unfolding developments with Flynn and the White House. The DOJ attorneys we were about to brief were career employees we had known for years and who had worked with us on an almost daily basis since the inception of the Crossfire Hurricane cases. They were as concerned as we were about the ever-expanding network of connections between Russia and the new administration. What's more, the senior levels of DOJ, not the FBI, bore the responsibility of interacting with the White House. Whatever solution lay ahead, DOJ had the unenviable role of navigating the space between that investigative world of the FBI and the political one of the White House.

About five people were in the room for the briefing when I started. I opened up with an explanation of how the interview came to be, then walked through the questioning itself, including Flynn's puzzling demeanor as he provided inconsistent answers to our questions. Then we turned to a discussion of potential criminal violations. An attorney with whom I had previously discussed the Logan Act — the law from 1799 that prohibits an unauthorized person from negotiating with a foreign government on behalf of the U.S. — reiterated his view that no reasonable prosecutor would use that law to bring a charge given the facts as we then understood them. We also discussed whether or not Flynn might have made false statements to us. We all reached the conclusion that we weren't at a decision point yet but needed to gather more information, such as his phone records and those of others he might have spoken with on the transition team.

I hadn't noticed that the most senior member of the staff of deputy attorney general Sally Yates had quietly entered the room during the briefing and sat without speaking at the head of the table. Although assigned to the DAG's office, he had remained a de facto adviser to Yates, who had been elevated to the role of acting attorney general after Lynch had departed and prior to the arrival of Sessions, who at this point had not yet been confirmed.

Why is he *here?* I thought after I registered the DAG staffer's silent presence. It was his right to attend almost any meeting he wanted to at DOJ, but his arrival immediately signified a high-level interest that wouldn't be satisfied by later hearing from the other DOJ attorneys in attendance. I had no objection to his being there, but there's a by-the-books process that's typically set in motion before such a high-ranking official steps into the briefing sequence. A feature of bureaucratic government meetings is that both sides almost always have a symmetry of rank, and this suddenly unbalanced the scale. His unannounced arrival, leapfrogging that well-established bureaucratic process, immediately made me wary. I was glad that Baker, our general counsel, was there as I continued the briefing.

There was a consensus between DOJ and the FBI that we had reached the point when we needed to inform the White House about developments concerning Flynn. We wrapped up the meeting and agreed that we'd take the decision up the chain. As people began to gather their notes and pack up, the senior official from the DAG's office spoke up for the first time.

I also want to make sure we agree that the FBI will not conduct any overt investigation on this without approval from DOJ, he said. Everyone in the room heard it and, without trying to be too obvious about it, turned to look at him. It felt like the metaphorical needle being yanked off a record with a screech. It wasn't that he was wrong — the working practice is for the FBI to clear any investigative activity at the White House, including anything from an interview to a briefing to collecting evidence, with DOJ in advance — but the appropriate audience for his comment was Comey or McCabe, not us. After an almost imperceptible half-beat, Baker responded, *I'll take that request back to the director.*

The response was immediate. *This isn't a request,* the DAG staff member snapped. *The FBI will not conduct any overt investigation without going through DOJ.* Everyone stopped closing up and sat back down, eyes turning to the two men moving down a path of confrontation.

Like I said, I'll take that request back to the director, Baker reiterated.

Growing agitated, the DOJ official shot back words to the effect of, *If we need to have the attorney general order the FBI to take no further investigation without our approval, we will.*

Baker, though his mind must have been going a thousand miles an hour, remained calm, implacable. *The department is free to do whatever it feels is appropriate. But I answer to the director of the FBI, and I will take the request back to him.*

They were at an impasse. DOJ was undoubtedly upset that Comey had given us the green light to interview Flynn without coordinating with DOJ, and the FBI was trying to chart an investigative path that was beset by politics on all sides. The conversation ended. *Well, that went great,* I thought.

The venue for the argument was completely inappropriate. If the senior levels of DOJ wanted to have that discussion, they needed to have it with Comey and McCabe, not with lower-level senior executives like us. Everyone connected to Crossfire was under a lot of pressure, but we generally kept an even keel. This was one of the rare moments when the burden of the investigation burst into the open, and my sense was that everyone there understood the confrontation in the context of the extraordinary moment in which we found ourselves and the gravity of the case before us. I don't think anyone at DOJ — at least not yet — was trying to improperly influence us or protect the White House from appropriate review. I saw it more as a sign that people's nerves were beginning to fray, and took it as one of the many small ways in which the Trump administration's behavior was already beginning to chip away at the norms of past behavior of the FBI and DOJ.

The executive from the DAG's office departed, dispelling some of the tension in the room. The rest of us shook hands and discussed last-minute decisions about who was following up with whom. A feeling of relief that the confrontation had ended settled over the room, a subconscious reassurance that we were all okay with each other.

We filed through the turnstiles at the DOJ exit on 10th Street NW and stepped into bright afternoon sunlight. We turned north toward Pennsylvania Avenue and the crosswalk that led from DOJ to the FBI. As we approached the FBI entrance, Baker motioned for the rest of us to go in without him. *I'm going to walk around the block for a bit,* our general counsel said. And then he continued down the sidewalk and disappeared into the pedestrian traffic, lost in thought.

A VISIT IN CHICAGO

On the morning of January 27, two days after the contentious briefing at DOJ, George Papadopoulos's cell phone rang while he was in his mother's house in Chicago, where he was living. He glanced at the number and didn't recognize it but picked up anyway. A special agent from the Chicago field office, who was standing outside the house with a second agent, was on the line. They wanted to talk to Papadopoulos, but at the FBI field office. Papadopoulos put on a suit, got in their car, and went downtown.

I had given the green light to the agents to interview Papadopoulos. After the tense confrontation with DOJ on January 25, I had not seen — nor would I ever see — any evidence that DOJ issued the edict that the DAG executive had demanded. However, the need to resolve the potential ongoing threat to national security loomed large. Just as the urgency to resolve the Flynn investigation had increased since he was in such a sensitive government position, we were determined to try to conclude the other cases where the subject, like Papadopoulos, might be heading to a government position. Even if there was friction at the top of the FBI and DOJ org charts, the collegial relationship between the middle and lower levels of DOJ and the FBI remained as strong as ever from my perspective. Working together, we began taking steps to bring cases to a conclusion where we could, reaching out to interview not only people like Papadopoulos but others who might provide information to fill in gaps in our knowledge.

Interviewing Papadopoulos was part of that effort, but it was also a big new step for the Crossfire Hurricane investigation. For one thing, it brought

a risk of public exposure to the investigation, which we hadn't had to con-
tend with when interviewing Flynn. While Flynn's interactions with Kislyak
were now common knowledge, Papadopoulos's encounters related to the
Russians were not. No one in the public knew yet about the Russian offer
of assistance to the Trump campaign that Papadopoulos had relayed to the
allied foreign nation the year before.

Our investigative interest in Papadopoulos also was different from our
interest in Flynn. At the same time that we were trying to get to the bottom
of the Flynn-Kislyak interaction, we also needed to answer the unrelated
question of whether or not the Trump campaign had taken the Russians up
on their offer of election assistance. Papadopoulos, who was the first person
we knew of on the Trump team to have been aware of that offer, might hold
the key.

Papadopolous's conduct during the interview did not reflect well on the
former campaign adviser, who our investigation had revealed had continued to try to monetize and exploit his ties to the administration after leaving the campaign. He lied repeatedly to the two agents, a crime he would
later plead guilty to and go to jail for. He lied about the nature and extent of
his meetings with the Maltese professor Joseph Mifsud, as well as about the
Russians that Mifsud had introduced him to. He lied about the importance
of Mifsud and the role he had played in setting up contact between Papadopoulos and individuals connected to Russia. And while Papadopoulos
did admit that Mifsud had told him the Russians had dirt on Clinton in the
form of "thousands of emails," he lied about the date that Mifsud told him
that, claiming it was before he joined the Trump campaign.

When I heard the results of the Papadopoulos interview, I thought three
things. First, by lying to the FBI, Papadopoulos had just broken the law
and made himself look even more suspicious. We knew he wasn't telling us
the truth about things that we already had evidence of, which only raised
questions about whether there were things we *didn't* know of that he might
be concealing from us.

Second, regardless of the timing of his conversations with Mifsud, Papadopoulos confirmed that he knew about the Russians' possession of material damaging to Clinton before the U.S. government knew about it. Either Papadopoulos could see into the future to predict the release of the hacked DNC email or someone with privileged knowledge of Russian intelligence had told him it was coming.

Third, Papadopoulos confirmed the nature of the information held by Russia. While the original allegation from the allied nation said merely that Papadopoulos had claimed the Russian government had *information* damaging to Clinton, Papadopoulos told the agents that the information was in fact *email* — a revelation that proved he had privileged knowledge about the precise form of *kompromat* (email, and not some other type of information, such as intercepted phone calls or medical records) that Russia had offered to weaponize.

In some ways the Papadopoulos interview went well — for us. Our investigation gained valuable insights into the extent and nature of the engagement between the Russians and the Trump campaign. And we had made these gains without exposing our investigation of Papadopoulos or the Russian offer of assistance to the Trump campaign, the latter of which we were especially attempting to keep secret.

But in other ways the interview left big gaps in our understanding of the Crossfire Hurricane investigation. Prime among them: Had the Trump campaign accepted the Russians' offer of election assistance? Were members of the Trump campaign complicit in a foreign attack on our electoral system, or had they merely been the unwitting beneficiary of it? This was a question we still needed to answer, and we would need to look beyond Papadopoulos to do so.

FLYNN'S END

Flynn's case, like Papadopoulos's, was approaching its zenith. Discussion between DOJ and the White House about the national security advisor percolated into early February. Late in the afternoon of Friday, February

10, Priestap joined McCabe for a routine counterintelligence briefing of the vice president's staff at the White House. As I sat at my desk trying to clear up bureaucratic paperwork, my phone rang. It was Priestap, calling from his cell phone, his voice almost imperceptibly clipped and more precise than usual. *Pete, I'm on my way back from the White House. Would you be able to get someone to quickly pull the raw material we have on Flynn for me?*

Of course, I answered. I had the material in an increasingly dog-eared file locked in my office safe. *How soon do you need it?*

As soon as you can get it, Bill answered. *I need to take it back to the White House for the vice president. Andy's waiting there with him.*

At the moment that I was gathering the Flynn material to deliver to the White House, agents across town were going through their notes from a hotel lobby interview they had conducted earlier that day with Joseph Mifsud. We had discovered that he was in Washington, D.C., and had swooped in to surprise him. Mifsud told the two agents that he had no idea that Russia had email that would be damaging to Clinton. The professor claimed that he and Papadopoulos had only talked about cybersecurity issues and that Papadopoulos must have misunderstood him. Mifsud appeared to be lying, but it was hard to call him on it, because Papadopoulos had also lied to us about his interactions with Mifsud, telling us that he learned about Mifsud's claims of stolen email before he joined the Trump campaign while minimizing the importance and frequency of the contacts between him and the professor. We didn't know what the truth was, and bluffing is difficult when you're not sure what cards you're holding; we had no idea what weaknesses he might detect in our questions. After talking with the agents, he slipped away and left the U.S., never to return.

Meanwhile, I pulled my file of the Flynn intelligence information, made a quick copy, and was about to call to arrange a handoff when Bill appeared at my door. His face was drawn and tense; I could sense his stress. He told me that Pence, after initially standing up for Flynn on *Face the Nation,* was now worried that Flynn had lied to him about discussing sanctions with Kislyak. He wanted to review exactly what we knew Flynn had done.

This came as a significant relief. Pence hadn't been knowingly covering for Flynn. The vice president wanted to preserve his credibility and reputation, and the White House appeared to be willing to go to the mat on his behalf.

Bill locked the intelligence into a courier bag and walked out to his waiting car for the ride back to the White House. As soon as he left, I knew with certainty that Flynn's days were numbered. If I was right about Pence, then it was a safe bet that he'd be livid once he saw how badly Flynn had deceived him. I wasn't even sure Flynn would make it through the weekend.

He did, but barely. On the evening of Monday, February 13, Flynn resigned. He noted in his resignation letter that he had "inadvertently briefed the Vice President Elect and others with incomplete information regarding [his] phone calls with the Russian Ambassador." As I read his letter, I wondered whether Flynn would claim he had also "inadvertently" told us "incomplete information" during our interview. I couldn't square all the claims of forgetful inadvertence with the untruths of Flynn's statements to all of us.

It also struck me that Flynn's public explanation for his resignation must have been a very bitter pill for him to swallow. If he was going to claim, in the face of very clear evidence to the contrary, that he had repeatedly and inadvertently failed to tell all of us these extraordinarily notable and relevant details about his conversations with Kislyak, then he was placing his innocence above his competence. But in either case, a lie is a lie.

"LET THIS GO"

Pence had demonstrated rectitude in his handling of the Flynn imbroglio, but that virtue didn't appear to extend to Trump. No sooner had we breathed a sigh of relief over Flynn's departure from the White House than we were presented with fresh cause for concern. This time the source was the man whom Flynn had admired so much — the chief executive of the nation, who we knew had his own problematic relationship with Russia.

The day after Flynn's resignation, Comey went to the White House for a routine counterterrorism briefing in the Oval Office that has since become famous. A half-dozen people attending the classified briefing arrayed themselves around the president at the Resolute desk, including newly installed attorney general Jeff Sessions. At the end of the briefing, Trump asked Comey to stay behind to talk.

Once everyone else had left the room, Trump told Comey, "I hope you can see your way clear to letting this go, to letting Flynn go. He is a good guy. I hope you can let this go."

I didn't know it at the time, but Trump's request was part of a series of one-on-one interactions Trump had with Comey in which Trump appeared to be pressuring Comey to interfere with our investigation of Flynn and, more broadly, in which the president tried to get Comey to affirmatively clear Trump or anyone related to him of any wrongdoing relating to Russia. Trump also asked Comey for an oath of loyalty to the president, a request that Comey took to be a demand and tried to sidestep.

Unbeknown to Trump, Comey had been carefully documenting these episodes in a series of confidential memos, which would soon become widely known and would contribute to the decision to considerably expand the scope of the investigation into the Trump team's contacts with Russia. But on that Valentine's Day, the interactions between Comey and the president were still a closely guarded secret, known only to a tight circle of people around the FBI director, including McCabe and Jim Baker.

I heard some oblique references to Comey's interactions with Trump, but it wasn't until at least several weeks later that I found out the full extent of their back-and-forth. One midspring morning Comey arrived late for a briefing that I was attending, apologizing for his tardiness as he sat down. When the main meeting ended and the briefing group skinnied down for a Crossfire update, Comey explained that he had received an odd, out-of-the-blue phone call from Trump, who had told him that he was just checking in to see how he was doing. Describing what was apparent to me to be more than just one phone call, Comey mused about the pattern of the president's behavior, what Trump was trying to achieve, how the president thought about it, and what the FBI should be doing in response. I listened, appalled but not completely surprised. That response then morphed into being appalled that I was *not* surprised. I remember thinking that it looked like Trump was trying to pressure the FBI to influence an investigation related to his administration, one in which he was quite possibly implicated. Now he appeared to be trying to obstruct it. It was nothing that I had ever thought I would see from an American president.

Trite as it might sound, I also felt a wave of gratitude for Comey's ethical and moral leadership, and for his almost pugilistic willingness to take on Trump, who technically could fire Comey anytime he wanted. Normally, none of us would have seriously considered that possibility, I think it's safe to say, because we couldn't imagine that the president would be reckless enough to fire one of his own intelligence chiefs while that chief's agency was investigating his employees and associates. But Trump was different, and the fact that Comey was willing to put his job on the line by standing up to the president both reassured me and gave me and others some ill-defined fear about Comey's job security. At least somewhere in Washington there were still leaders with values beyond their own self-interest.

When the meeting ended, we rose to our feet. The room was always eerily quiet, the carpet and walls swallowing the ordinary scuffle of feet and din of meetings inside it, but after this meeting it was even more hushed than usual. As I prepared to leave, Baker took my arm and pulled me aside.

How do you feel, hearing that? he asked.

I was quiet for a moment as I thought about it. *It makes me angry. It's wildly inappropriate. I think he's —* Trump's *— gone further past the boundary of the rule of law than any president I know.*

But it's not going to impact your view of the investigations? Baker asked.

Oh — no, I responded. *It doesn't deter me and it doesn't change what we're doing. I'm okay.*

Baker nodded. Then he explained that although Comey had clearly just decided otherwise, the boss had previously decided not to talk about his interactions with Trump with the Crossfire team because he did not want that knowledge to inappropriately influence investigators. If Trump, as the chief executive, wanted to limit the FBI's investigation, that might influence the investigators. It might make them scared; it might make them angry. It could do a number of things that could threaten the independence of those investigations. I told Baker that I understood the reasoning. I thanked him and reassured him again that he didn't need to worry about us. While I appreciated Comey's and Baker's concern, I also thought it was unnecessary. We were all professionals; the team was going to follow the investigation wherever it led, to innocence or guilt, regardless of what Trump said.

Driving home that evening, I wondered whether Comey's conversations with Trump had crossed the line to the point where we needed to finally take the step I had been arguing against for weeks: opening an investigation of the president himself. Arguably, Trump's interactions with Comey weren't about Russia but rather about obstructing the FBI's investigations into Trump's associates — albeit in a way that ultimately related to and benefited Russia. After all, if Comey acceded to Trump's demands by halting the Bureau's investigation of Flynn or downplaying the suspiciousness of the connections between the Trump administration and Russia, the result would be that this invisible web, and the influence it implied, would remain intact.

This simple fact was what began to sway me, that night, toward supporting an investigation of the president. Trump's obstructionism wasn't just good for him and his cronies; it was also good for Moscow.

But I trusted Comey's intuition. I was confident that he would recognize when and if we had arrived at the point at which we needed to open an investigation on the president himself, beyond those we already had opened on a number of people who had previously worked for Trump. In the meantime, I honored Comey's wishes and kept mum about his discussions with Trump, letting the team pursue their various investigative avenues without the burden of that extra information.

But trying to keep that knowledge quiet wouldn't matter for long, as I found out when I walked into the Counterintelligence Division front office about a month later. It was early evening on May 9, and Bill Priestap's assistant was at her desk, eyes glued to a large-screen TV tuned to CNN. The network was displaying live aerial video of a black SUV driving on a busy freeway in Los Angeles. Though I had no idea what I was looking at, I immediately thought of O. J. Simpson's white Ford Bronco and his slow-motion flight through L.A. traffic in 1994.

What happened? I asked.

Trump just fired Comey, came the answer.

PART III

12

Sentinel

TO BE A successful FBI agent, you must be able to impassively absorb unexpected and traumatic information. That instinct kicked in as I stared at the screen outside Bill's office on the evening of May 9 and struggled to maintain a poker face. While I intellectually grasped what was happening, I had difficulty processing it. After a moment I walked into my office, grabbed a Coke Zero out of the small refrigerator in the corner, and sat down at my desk to watch the news for a few minutes. I gathered my thoughts as I watched the overhead helicopter coverage of Comey's motorcade snaking toward the airport. Then I walked across the front office's reception area to Priestap's office to take our bearings.

The news of Comey's firing ricocheted through headquarters. Even though it was after hours, I could sense the apprehension as we hurriedly convened to discuss Comey's sudden termination. Agents gathered in doorways, watching the unfolding event on TVs in offices that should have been closed and dark at that hour. A palpable sense of disbelief hung in the hallways.

It wasn't the fact of his firing alone that was so shocking; it was how it had taken place. Comey had flown to Los Angeles for a recruitment event intended to bolster the ranks of minority agents in the Bureau. Before the event, he had visited the L.A. field office. He was in the middle of a speech to the L.A. employees, driving home the mission to protect Americans and uphold the Constitution. As he spoke, the TV screens on the back wall behind the audience displayed "Comey Resigns." He laughed and made a joke, and then the message changed to "Comey Fired." The buzz in the room rose. Comey cut short his remarks, shook hands with the assembled employees, and then retreated to a back room to learn, not from a phone call from the White House or the attorney general but from a cable news network, that the president of the United States had fired him.

That evening the White House issued a press release containing three documents. One was a memo from deputy attorney general Rod Rosenstein addressed to Attorney General Sessions. In it, Rosenstein criticized Comey's decision to announce the previous July 5 that the Clinton email investigation had ended and that no prosecution was warranted. "The Director announced his own conclusions about the nation's most sensitive criminal investigation," Rosenstein wrote, "without the authorization of duly appointed Justice Department leaders." What's more, while he stopped short of recommending Comey's removal, Rosenstein seemed to imply that it was necessary: "Although the President has the power to remove an FBI director, the decision should not be taken lightly. I agree with the nearly unanimous opinions of former Department officials. The way the Director handled the conclusion of the email investigation was wrong. As a result, the FBI is unlikely to regain public and congressional trust until it has a Director who understands the gravity of the mistakes and pledges never to repeat them. Having refused to admit his errors, the Director cannot be expected to implement the necessary corrective actions. "

The second document was a letter from Attorney General Sessions to President Trump, presenting him with Rosenstein's memo and explaining Sessions's conclusion that "a fresh start is needed at the leadership of the FBI."

The third document in the White House press release was a letter from Trump to Comey, announcing his termination on the basis of Rosenstein's and Comey's complaints. It wasn't the original, several-pages-long letter that Trump had written from his Bedminster golf course (a letter that I would eventually see a few weeks later); rather, it was an anodyne replacement version pushed on Trump by lawyers. "Dear Director Comey," Trump wrote,

> I have received the attached letters from the Attorney General and Deputy Attorney General of the United States recommending your dismissal as the Director of the Federal Bureau of Investigation. I have accepted their recommendation and you are hereby terminated and removed from office, effective immediately.
>
> While I greatly appreciate you informing me, on three separate occasions, that I am not under investigation, I nevertheless concur with the judgment of the Department of Justice that you are not able to effectively lead the Bureau.
>
> It is essential that we find new leadership for the FBI that restores public trust and confidence in its vital law enforcement mission.
>
> I wish you the best of luck in your future endeavors.

I drove home that night feeling profoundly unsettled. Over the past five months my unease about the administration's response to and interaction with Russia had steadily turned to alarm. Now it seemed that Trump had knocked still another keystone out of the rule of law. I felt that the president had just unleashed a broadside against his own government, one that compared with Nixon's attacks on his own Justice Department in the darkest days of that presidency. But Nixon's MO largely had been to preserve himself. Something even more corrosive, not merely self-interest, seemed to be at work in this administration.

I ran through the fact pattern in my head. The puzzling, unexplainable loyalties to Putin; an attorney general who had misled lawmakers about his campaign contacts with Russia; the newly installed deputy attorney general, Rod Rosenstein, who had written a letter critical of Comey's public handling of the Midyear investigation that the White House clumsily wielded to justify Comey's firing; and on and on. The Trump administration's actions vis-à-vis Russia were highly suspicious, highly consistent, and highly advantageous to America's historic adversary without clearly benefiting, and at times even disadvantaging, our own security and stability.

Now Trump had effectively decapitated the federal agency responsible for scrutinizing such suspicious behavior. It seemed obvious to me — and to most people at the Bureau at that time — that the president may have obstructed justice by firing Comey. His apparent obstruction was even more blatant than when he had pressured Comey to halt the investigation of Flynn. And this time he had not merely *attempted* to effect a change at the FBI that would help Russia; he appeared to have succeeded. From a counterintelligence perspective, there was too much smoke for us not to continue to look for a fire. And the closer we got to the Oval Office, the stronger the smell seemed to become.

After I went to bed on May 9, I lay awake for hours, unable to sleep. Finally I slipped out from under the covers and tiptoed downstairs. I turned on the lights in the kitchen and sat at the table. With tired eyes I looked down at the tabletop, where the soft wood bore the indentations of years of the kids' postdinner homework. Then I stared out the back window into the darkness of the backyard, deep in thought.

Were we overestimating the gravity of the situation? I wondered to myself. *What else should we be doing to protect the investigations? Who could we trust?* And beneath all those concerns, the fundamental question, which I had argued against for months: *Had Trump passed the threshold for us to open an investigation of him?* In the stillness of early morning, I didn't find any answers.

After an hour or so I fumbled my way back into the darkened bedroom for a few fitful hours of rest. I hoped the new day might bring some reprieve from the flood of bad news. But after months of grim revelations, I had a feeling that things would only get worse.

"GREAT PRESSURE"

When I returned to the Bureau the next morning, a sense of collective shock hung over headquarters. Comey's firing felt like a sudden and unexpected death, one that could only be whispered about behind closed doors. On

TVs playing around the building, deputy press secretary Sarah Huckabee Sanders claimed that the White House had heard from "countless members of the FBI" who didn't support Comey. It was an assertion that she later described to the special counsel as a "slip of the tongue," but I knew it was untrue when I heard it. The FBI, and headquarters in particular, is one of the least emotive places I know, but I saw more expressions of mutual support among the workforce that day than I had seen since 9/11. The mood at the J. Edgar Hoover Building was somber, and the White House's inaccurate spin to the press corps about it only worsened the outlook.

At some point after the firing, I went up to Comey's office to pick up or drop off something. While I waited, I looked into his vacant conference room and noticed a row of brown boxes stacked along the wall: Comey's personal effects, which would be returned to him. Comey's assistant, an elegant woman whom I had known for years, came around from her desk and looked into the room with me. In my experience, she rarely betrayed any emotion other than an occasional circumspect smile. *How are you doing?* I asked. She hugged me, and the tears in her eyes provided the answer.

Comey's firing had a profound effect on Crossfire Hurricane, just as it did on the Bureau as a whole. Ever since details of our investigation had been shared with the Trump administration by Congress shortly after Comey briefed them in early March, the team had been careful to watch for signs of improper interference by political actors. Trump's pressure on the investigation, including his request to Comey to "go easy" on Flynn and otherwise put a premature end to the Bureau's Russia probe, had increasingly raised eyebrows, as well as our alert level, at the FBI. Comey's firing was like a perimeter flare in the night: the long-hoped-against assault might have just begun.

In the immediate aftermath of the director's termination, the memory of White House personnel destroying evidence following the discovery of the Watergate and Iran-Contra scandals hung in the air as we all considered how best to protect the U.S. from Russia. We worried that an order to destroy Comey's documents might come down from DOJ, or even the extreme scenario of a standoff as his papers were boxed up and important evidence was carried away by DOJ or others, possibly to be destroyed. Facing such prospects, we scrambled to preserve evidence and ensure that our investigation could not be illegally interfered with. Even that was subject to debate: could the president lawfully order the investigation to be closed? FBI lawyers told us there were arguments that he could.

Calls from Congress and the media for the investigation to be protected from interference began almost as soon as Comey's termination was announced. In fact, in time the evidence contained in Comey's records would give substance to suspicions outside the Bureau, as well as within it, that the president had obstructed justice by firing his FBI director. Comey's files also would offer proof that this was not an isolated act of obstructionism but rather an insidious pattern of behavior — one that benefited not just Trump but also Putin.

As I've mentioned, I had learned in the weeks before Comey's firing that Trump periodically contacted Comey for one-on-one talks and confidential discussions. I would soon see the memos that Comey had been typing up about those interactions, documenting what the president had been saying in case it was relevant to an investigation. He had written seven memos in all. Now that Comey was gone and the investigation was in the president's direct line of fire, we felt it was our duty to protect those memos and the evidence they potentially represented by preserving the words that the president had used behind closed doors. McCabe shared our desire to protect the memos from unlawful destruction, so, to preserve the records, he directed the Crossfire team to place copies that could not be expunged into Sentinel, the FBI's computer case file system.

That very same day, Trump himself was underscoring the importance of those memos and the investigation to which they would soon contribute. That Wednesday morning—the day after he fired Comey—Trump welcomed Russian foreign minister Sergey Lavrov and Ambassador Kislyak into the Oval Office. I learned about the meeting at the same time that the American public did, and not from U.S. media. The White House had barred American reporters and other media from the meeting — in the *U.S. president's office* — but allowed in Russian photographers. They released pictures of Trump in a red and silver repp tie, beaming at Kislyak, the president's right hand clasping Lavrov's arm. Tom Clancy or John le Carré could not have scripted fiction more alarming than what I was seeing.

I was disgusted by the sight of Trump glad-handing representatives of the government that had just attacked our electoral system. But when I found out later what Trump had said to them, my blood ran cold. Apparently in a jocular mood, Trump had announced to the senior Russian officials, "I just fired the head of the FBI. He was crazy, a real nut job." Then, from the office shared by Lincoln and Eisenhower and Reagan, Trump continued to his Russian visitors, "I faced great pressure because of Russia. That's taken off."

Our decision to protect the Comey memos by securing them in a classified database no longer seemed like such an extreme precaution. We had no way of knowing that on this very same day the White House faced a similar dilemma regarding the official record of Trump's comments in the Oval Office — and devised a similar solution, but for an altogether less noble purpose.

During his meeting with Lavrov and Kislyak, Trump not only crowed about firing Comey. As the *Washington Post* later reported, he also inexplicably told them he wasn't concerned about Russian interference in the 2016 election because the U.S. did the same sort of meddling in other countries. In one fell swoop, one burst of moral relativism, the American leader had absolved the Russians of their crime against our democracy. In his naked attempt to reassure the Russians that they wouldn't be held accountable for their interference in our democracy, Trump's remarks were right in line with Flynn's secret call to Kislyak about the Obama sanctions — except this time the message was delivered in front of Russian cameras on the carpet of the Oval Office.

Anyone around Trump in the White House who heard it and had any understanding of national security knew they had a problem. They decided they had to hide the memorandum documenting what was discussed in the meeting. According to the *Washington Post,* the White House immediately restricted an internal memorandum summarizing the meeting, keeping it from National Security Council (NSC) personnel who would ordinarily have access to it.

In securing the Comey memos in the FBI database, our goal could not have been more different from the White House's. We were trying to preserve evidence of the truth. They were trying to keep it from being found.

From a counterintelligence perspective, the Oval Office meeting was like a five-alarm fire. In absolving the Russians of their election meddling, Trump also made clear that he was unlikely to hold them to account for similar activity in the future. Moreover, after telling the world that Comey had been fired because of his handling of the Clinton investigation, Trump had just told the Russians the truth: he had fired Comey to put a stop to the Crossfire investigation.

Standing in the White House with the Russians, Trump appeared to compromise himself yet again. Although the FBI didn't learn about his remarks immediately, the Kremlin certainly did. Trump had tacitly blessed the Russians for their attack on our democracy while giving them a green light to continue. This inflammatory comment, delivered away from the prying eyes and ears of the American media and covered up afterward, likely gave the Russians yet another point of leverage over him.

Ironically, if Trump had publicized his comments, they wouldn't have been quite as bad. It was the attempt to conceal his remarks, much like his lie about Trump Tower Moscow and Flynn's secret request to Kislyak, that added to the growing coercive power that Moscow held over the American president. Recall that *kompromat* doesn't need to be acknowledged to be effective — sometimes it's more effective when it's not. Simply by being mutually aware of its existence, the subject and the possessor of the compromising material enter into a coercive relationship. The coercer does not need to give instructions to the coerced. This is what makes *kompromat* so pernicious — and so devastatingly effective. This is what made Trump's deception and hidden Russian communications so dangerous for the nation he had just been elected to lead.

"PETER, I'M STARTING TO HATE YOU"

I tried to be circumspect as May 10 ground on interminably. I repeated to myself, *We can't rush this investigation; we need to slow down the pace and be careful.* Hasty decisions, especially those made under pressure with incomplete information, carry risk. But it was a challenge to keep calm when events were mounting so quickly. And I knew that whatever we felt at our level, the stress on McCabe — now the Bureau's acting director in the wake of Comey's removal — must be greater still.

But try as we might, there could be no slowing down. Comey had been gone for less than a day, but already a number of events loomed, including a worldwide threat briefing to the Senate Intelligence Committee the next day that would be broadcast to the public. Comey had been planning to deliver it, but that task now fell to McCabe, who had less than a day to prepare.

None of us were concerned that McCabe, a counterterrorism expert, would have any problems with questions related to terrorism. But that was unlikely to be the focus of the hearing. We all knew it would be heavily focused instead on Comey's firing, Trump, and Russia. McCabe's remarks at the briefing would constitute the FBI's first public comments since Trump had announced Comey's termination on Tuesday night. Congress would demand answers. McCabe needed to be able to deliver.

That night a handful of us sat around the table in the deputy director's conference room to help McCabe prepare for the Senate threat assessment briefing in the morning. We felt dispirited as well as physically and intellectually exhausted. While I'm sure McCabe felt that too, for a man thrust into the hot seat, he seemed remarkably well collected. Wearing his retro horn-rimmed glasses, he glanced down at the open binders that encircled him on the tabletop as we peppered him with questions. He would jot an occasional note to himself on a pad of paper in front of him.

Thinking ahead to what McCabe might face in the Senate, we tried to trap him in his answers and rattle him with hostile, rapid-fire interrogation. As we took turns questioning him, after one particularly mean-spirited exchange with me, McCabe swiveled in his chair and said jokingly — mostly

— "Peter, I'm starting to hate you." *Good,* I thought. *With any luck, this will be worse than anything you face tomorrow.* In hindsight, this preparatory grilling was even tougher for McCabe than I could have imagined at the time. As I would learn later, it's one thing to prepare someone for a contentious congressional hearing, but it's an entirely different matter when you're the one getting ready to testify.

In McCabe's case, the most challenging question that night was how to characterize the feeling in the Bureau's workforce following Comey's termination, given what the White House had said. We offered options for how to discuss Comey's firing, whether to remain circumspect or to be more vocal and describe the strong and broad support Comey enjoyed within the FBI, which would contradict the description from the White House. We listed reasons for and against both approaches. We weighed the benefits of describing Comey's leadership, respect for him among the FBI's workforce, and how the Bureau's employees were taking his firing. We talked about how directly to take on or otherwise rebut the false characterizations coming out of the White House. McCabe absorbed our suggestions; it would be up to him to make the final call.

Despite the difficult circumstances, McCabe was easy to prepare. He was both a bright man and well versed in the internecine political warfare of inside-the-Beltway Washington. Sometimes when preparing someone, I would have to make a suggestion, explain my reasoning, then debate the merits. McCabe — like Comey — almost never needed the second step, immediately intuiting the reason behind the comment.

The hours went by, and it grew more difficult to focus. I could see the strain on the faces of the people in the room, and I could sense McCabe reaching a point of diminishing returns, becoming less patient with the occasional dumb suggestion. Like studying for a test, at some point it's better to close the book and think about the big picture. *Okay, folks, I think I'm good. Thanks for all your work,* he announced, closing up his binders to return to his office for what I was sure was a difficult night ahead.

It was well into the night when we left the prep session, but my work still wasn't done. There was more to prepare for the next day: separately from

McCabe's Senate hearing, we had a previously scheduled ICA briefing about the U.S. intelligence community's assessment of the Russian interference campaign for Rosenstein. Rosenstein's staff had just let us know that the briefing was still on, so I went back to my office to look through the material on my desk, going back to the beginning of Crossfire to reacquaint myself with names and dates and thumbing through timelines.

At some point late that night, my boss, Bill Priestap, received most of Comey's memos about Trump from the seventh floor, and I got the job of getting them into the Sentinel database. I wanted to read them, but the more immediate task was to incorporate them into the investigative file. The subfile we had chosen as the appropriate repository was so highly protected that even most of the Crossfire team didn't have access. But I did.

I got up from my desk, stood at the bank of printers between Bill's office and mine, and fed the memos into the multifunction printer used for secret documents. Once they were digitized, I could file them in Sentinel.

As a softly whirring motor pulled page after page into the scanner's document feeder, I stared out through the window blinds at the empty street below. The wide windows in front of me looked out over Pennsylvania Avenue, which was quiet and still but for the occasional taxi or car. On the far side of the street, across a line of concrete planters and low trees, rose the limestone walls of the Department of Justice. At the top was one of several engraved phrases adorning the building: "The Place of Justice Is a Hallowed Place."

Trying to imagine the mood within the Justice Department during Nixon's Saturday Night Massacre, I briefly wondered if we would be measured against that moment in history as I secured the Comey memos in the digital depths of our computer system, one after another.

"A MADE-UP STORY"

Hours later I was in that building across the street for the morning ICA briefing. I sat across the table from Rosenstein in a secure, windowless DOJ conference room, eyeing the man who had written the memo that Trump had used to justify firing Comey two days earlier.

Rosenstein is tall but not unusually so, trim and bespectacled, and has a quiet demeanor that, depending on the circumstances, can appear amiable or impassive. A George W. Bush appointee, he was a career prosecutor who had built a solid reputation and a record of success as a U.S. attorney in Maryland. I had worked directly with him only once, and generally viewed him as a straight shooter who pursued and won cases on merits, not on politics. But now, as I sat across from him, I wasn't so sure, and I had to suppress a deep sense of suspicion.

The White House had tried its best to put Comey's firing on Rosenstein. It "was all Rosenstein," press secretary Sean Spicer said the night of the firing, while Sarah Sanders added the next day that Trump "had accepted the recommendation of" the DAG to remove Comey. That seemed like an exaggeration — Rosenstein's memo, while highly critical of Comey and alluding to his removal, stopped short of recommending it — but having read all three documents, I couldn't contest that Rosenstein's memo to Sessions appeared to have provided Trump with the pretext he needed to fire Comey.

Now, as I tried to focus on the ICA on May 11, I had to fight to quell the urge to ask the question burning in the forefront of my mind: *Did you provide cover to the White House to legitimize Comey's firing?* And there was a second question, which applied regardless of Rosenstein's intentions: *How, as the DAG, could you have written that letter and not have expected that it would be used to justify Comey's firing?* In my mind, either he had willingly participated in papering a record to help the president fire the nation's top lawman, or he had been unable to recognize that he was going to be used for that purpose. Either scenario caused me to be on my guard.

I found out a few days later that both Rosenstein and his boss, Jeff Sessions, had independently called White House counsel Don McGahn to protest the way the White House portrayed the firing to the public and had asked Trump to correct the record to make clear that he had decided to fire Comey before Rosenstein wrote the memo. Needless to say, the president didn't do that, at least not immediately, although Trump's press officers at least began equivocating in their accounts of what had happened. The apparent surprise of DOJ leadership would remind me a bit of Flynn's wide-eyed astonishment at the personal attacks leveled against him, a sort of political naïveté about the highest levels of government, or at least the new Trump version of government. *How could you have watched the early days of the administration,* I wondered, *and not have expected this?*

That Thursday, I did not yet know Rosenstein's true role in Comey's firing, so I kept my guard up as I sat in the bowels of the Department of Justice, across the table from the deputy attorney general. I didn't have to speak first in the briefing, so I watched as colleagues from the CIA, NSA, and DNI walked Rosenstein and his staff through what the Russians had done and laid out the classified details of the attack. My turn came about midway through the briefing. I talked about the broad scope of what we had seen the Russians doing in the U.S. Rosenstein absorbed it all quietly, asking few questions.

As we finished and gathered our materials to leave, Rosenstein thanked the group. And then he asked, *Pete, would you stay behind for a bit?*

As a rule DAGs don't interact with FBI deputy assistant directors such as myself, but I reluctantly agreed. Not that I had much of a choice, since Rosenstein was far above me in the law enforcement hierarchy, and anyway I wanted to hear what he was going to say. Still, much like the DOJ meeting after the Flynn interview, I found myself wishing that McCabe, Baker, or someone else from the seventh floor was beside me.

After the doors closed behind the departing team, Rosenstein asked me about a matter that hadn't been in the briefing and that the overwhelming majority of the intelligence community didn't know about: the FBI's cases on individuals connected to the campaign and administration. While Comey had announced in an open hearing before Congress on March 20 that the FBI was "investigating the nature of any links between individuals associated with the Trump campaign and the Russian government," he hadn't said who, how many, or what had led us to do so. From Rosenstein's questions, I got the sense that he had already heard some detail about them from his staff. It was his prerogative to ask, but alarms rang in my head when he did. Why did he want this information, exactly, and what was he planning to do with it? But he had the right to it, so of course I answered.

We sat down at the briefing table again, and I began laying out the investigations so far. Two members of his staff who had already been briefed stayed behind to listen. As I walked through each of the cases, Comey's firing loomed in the back of my mind. Rosenstein listened, occasionally asking questions but outwardly providing no hints as to his thoughts or opinion. That is, until we arrived at the final case on my list.

After ticking through the rogues' gallery of Papadopoulos, Manafort, Flynn, et al., our discussion turned to an investigation we had just opened: a perjury case on Rosenstein's boss, Attorney General Sessions.

The prospect of opening such a case had been percolating for weeks before Rosenstein had assumed office. Almost three weeks earlier, on April 21, McCabe and Baker had met with a senior member of the DAG's office to discuss the perjury referral sent by the Senate following Sessions's assertion during his confirmation hearing that he had had no campaign interaction with any Russians. The *Washington Post* had subsequently reported that Sessions had spoken with Kislyak twice in the months leading up to the election, during which time the Russians were vigorously attacking the country's electoral system. On March 2, the day after that news broke, Sessions had recused himself from any involvement in the ongoing Russia investigation.

During their meeting, McCabe and Baker had told DOJ about the information we had confirming that Sessions had had contacts with Kislyak that he had not disclosed. DOJ responded by asking the FBI to wait until Rosenstein was confirmed and then seek his input, so we did. After Rosenstein's confirmation on April 25, Comey personally broached the topic with him twice: once on April 28 and again on May 1, just eight days before Trump fired him, asking for concurrence to open an investigation. Rosenstein didn't take a position then, and he didn't now, sitting in front of me.

I worried that Rosenstein, and DOJ generally, had been trying to slow-roll the issue by delaying a decision until it simply went away. Whatever the reason, the delay just added to our suspicion about Sessions and the new leadership of DOJ. The facts supported opening a case, and any reason for delay was not based on a lack of legal authority to do so. Neither I nor anyone I knew in the FBI was convinced that we were seeing an enormous conspiracy playing out, but we also couldn't rule it out. As ever, we needed to consider and prepare for the worst.

What's more, in my mind we were on firm ground for opening a perjury case on Sessions. The facts and precedent were clear on this point; there was no question that the allegations of Sessions's lying put us well past the legal threshold for triggering an investigation. With Comey's dismissal and McCabe's elevation as the FBI's acting director — who knew for how long — I felt strongly that we needed to act swiftly rather than wait to see what the White House and DOJ did next. I worried that while we were seeking the simple truth to protect national security — had the attorney general deliberately concealed his contacts with Russia while part of the Trump campaign? — McCabe might be replaced by a new FBI director who wouldn't share his independent, truth-seeking spirit. We had delayed long enough.

As we talked, Rosenstein voiced some skepticism about our information regarding Sessions, offering mitigating interpretations about some of it. His objections were reasonable, but they weren't enough to justify standing down from investigating. Privately I suspected it would be neither a long nor a difficult investigation and might consist of little more than a check of our records and a single interview with Sessions.

I also sensed, as I believe many of us did, that this discussion about Sessions came against the backdrop of the elephant in the room: whether or not to open an investigation of Trump. If Sessions was concealing and lying about his contact with Russia, his behavior would fit a pattern by now well established by Papadopoulos, Flynn, and others in Trump's orbit. All of which led inexorably to a fundamental question: was Trump doing the same thing?

That question loomed large in my mind, but neither Rosenstein nor I mentioned Trump in that moment. Instead we stayed focused on Rosenstein's boss and our investigation into the perjury allegations against Sessions. Finally our discussion concluded, and I left the room feeling unsettled, as Rosenstein stayed behind to continue the conversation with his staff.

Later that evening Trump sat for an interview with CBS anchor Lester Holt. I didn't watch it when it first aired, but was dumbfounded when the press began reporting about it later that night. I immediately pulled up a replay online. When Holt asked the inevitable question about Comey, Trump's response was startlingly candid. Despite what the White House press office had been saying, Trump told Holt that Comey's firing had nothing to do with Rosenstein's letter. Instead it had been all about the Russia investigation. Our investigation.

Ten months into a sprawling investigation, with fresh memories of the previous day's Oval Office meeting between Trump and Lavrov, I listened as Trump explained to Holt: "Regardless of recommendation, I was going to fire Comey knowing there was no good time to do it. And in fact, when I decided to just do it, I said to myself — I said, you know, this Russia thing with Trump and Russia is a made-up story," concluding with what seemed to be a telling admission: "It's an excuse by the Democrats for having lost an election that they should've won."

Although characteristically shambolic, Trump's statement struck me as both the truth and an odd strategy: admit the wrongdoing boldly, then barrel through it as if it weren't wrong after all. But I didn't know just how strange this episode was until later. To the small circle of people who knew about Trump's comments to Kislyak and Lavrov in the Oval Office, the president had just demonstrated his willingness to tell the Russians one thing and — at least for a while — the American people another. On this point, at least, Russia had enjoyed its leverage over Trump for precisely one day: from his May 10 meeting with Kislyak and Lavrov to his May 11 interview with Lester Holt. Trump's cover story for firing Comey had held up only a day longer.

When news about Trump's comments to the two Russians leaked out in the press a little over a week later, I retraced the short life of this particular bit of *kompromat* with dismay and no small amount of disbelief. It was tempting to seek reassurance in the president's clear inability to keep this particular secret, or at least his lack of interest in doing so. But just because he let slip the real reason for Comey's firing didn't mean that there weren't plenty of other skeletons in his closet that he'd prefer to keep hidden. The question was how we would be able to protect the U.S. — and Trump himself — from the consequences if the Russians found them.

13

The Decision

CROSSFIRE CONTINUED TO move forward, but as we approached the end of the tumultuous week of May 8, it was clear that we were beset on all sides. The Russians and the White House kept up their attacks on us, continuing to attempt to upend our work. Now we weren't even sure if the DOJ was on our team. I felt as though we were barely hanging on.

To add to the pressure, we now were forced to confront two critical questions that had been steadily building for months. The first was whether to investigate the president himself and determine the nature of his relationship with the Russian government. The second was whether a special counsel should do it.

Trump kicked off Friday, May 12, with an early-morning tweet announcing, "James Comey better hope there are no 'tapes' of our conversations before he starts leaking to the press!" That later prompted Comey's memorable rejoinder to Congress: "Lordy, I hope there are tapes." The exchange highlighted the position we were now in: the firing of the head of the FBI by the president, which could be a possible effort to obstruct the investigation of connections between his campaign and Russia, added a new violation to an investigation overseen by a Department of Justice with an attorney general who had recused himself because of his key role in that campaign. It was a mess of conflicting motivations that created the real potential for improper interference.

Internal debate about the need for a special counsel — an outside attorney charged with taking over cases while reporting directly to the attorney general — had been raging within the FBI even before Comey had been fired. Those who supported the idea argued that the appointment of a special counsel would help to preserve the independence and integrity of an investigation that Trump increasingly seemed to be trying to undermine. A special counsel would also allow the FBI to return to a focus on the vast array of threats it faced every day without the political overlay of an investigation involving the White House.

Though the decision would be made well above my level, I felt that a special counsel offered an ideal path forward: it would convey objective independence, and it would shield the FBI and DOJ from the political blows that were being thrown at them by Trump on a routine basis. The week's events hardened the view — both my own and that of others within the Bureau — that DOJ should appoint one.

For some time McCabe had felt that a special counsel was inevitable. Comey's firing, combined with Trump's stated objective of getting rid of the Russia investigation, had cemented his position. With the president appearing to be committed to what many people in the FBI and beyond were coming to see as obstruction of justice, appointing a special counsel to lead the investigation now seemed like a necessary safeguard.

Late Friday afternoon, McCabe met alone with Rosenstein to push the DAG to appoint a special counsel to take over the FBI's Russia investigation. At times Rosenstein seemed out of his depth. At least twice, I later learned, he expressed regret that he hadn't been able to discuss the appointment with Comey, the man he had just had a hand in firing. I was shocked to find out that he asked McCabe if he was in touch with Comey, and if so, whether McCabe would pass on a message that Rosenstein was interested in Comey's thoughts on the matter. My surprise wasn't that Rosenstein thought Comey would have great insight — undoubtedly he would — but rather that Rosenstein thought it was a legally and politically sound idea to ask the FBI's acting director that question and to seek guidance from a fired official.

From the moment I opened Crossfire, I knew there was the possibility that we would reach the point at which we might be investigating Trump. We all did, while hoping we would never get there. After the election Baker had advised Comey to avoid saying that Trump wasn't under investigation — not because it wasn't accurate at the time, Baker argued, but because one day that might no longer be true. That would have placed Comey in the difficult position of needing to correct the record if a case was in fact opened, something we had learned the hard way from Midyear.

Needless to say, this was no longer Comey's problem; but it was ours. After the meeting in Bill's office where we had debated the counterintelligence issues of all the names associated with Trump that had been written on the whiteboard, we had occasionally returned throughout the spring to a consideration of the merits and appropriateness of an investigation of Trump.

For a long time I had argued against opening a case on Trump himself. It was a thorny, complex issue. I had tactical concerns about whether we would be able to quietly do any relevant investigation, and strategic and policy ones about the legal scope of our authorities and their outcomes. I remember Baker writing out a complicated grid on his whiteboard at one point, mapping legal options for a range of factors, such as whether an alleged crime was committed before or after the president took office, along with the legal authorities available at each stage through Article I of the Constitution (Congress) and Article III (the Judiciary). There were so many variables that we needed a visual roadmap to keep them all straight.

I hadn't wanted to investigate the president of the United States. But my conviction on that point had been eroded by Trump's continued suspicious behavior with the Russians and his ongoing attacks on our investigation. For me, as for McCabe and many of the FBI's remaining senior leaders, Comey's firing was the tipping point.

Now the extraordinary moment had arrived. By Monday morning, May 15, McCabe was close to a decision on whether the FBI would open a case on Trump. The discussion had shifted from whether we should open a case to whether there were any compelling arguments *against* it. At 11:30 a.m., a question came down to us from the seventh floor: was there any reason *not* to open an investigation of Trump? If not, we were told to open one within 24 hours.

McCabe deliberated overnight with all the levels of the FBI's leadership and counsel, from the director's level down to mine, and ultimately we concluded that there weren't any reasons not to commence an investigation of the president. We opened a case the following day, on May 16. The team hammered out the wording of the document that would serve to open the investigation and sent it via Baker for legal review and then to Priestap for his approval.

In some ways, reaching the decision to investigate Trump came as a relief. As nerve-racking as it was, making the decision was easier than having to constantly revisit the looming question as the factual predicate evolved. This is not to say that we took this step lightly. Nothing could be further from the truth.

Although I've said it before, I'll repeat it here as plainly as I can: from McCabe down to the Crossfire team, everyone who had a hand in opening an investigation into the president was acting with a clearheaded purpose of upholding the Constitution and protecting the American people. No one had hoped or wished for this. No one was happy. No one celebrated. No one questioned it. We had avoided it as long as we could. So, as we had so many times before, we gritted our teeth and got on with our work.

Once McCabe made the decision to open the case on Trump, a tangled knot of questions — legal and logistical — needed answering, some of which required Jim Baker and his attorneys to weigh in. Should obstruction be part of the investigation? Could we investigate the president for a crime that he couldn't be charged with while in office? Could he be charged while in office? There was a DOJ policy — never tested in court — that he couldn't. Would the focus of the investigation be more criminal or counterintelligence?

Some of these questions we could answer easily enough. For instance, because of a range of factors, from Trump's private request to Comey for the

FBI to end its investigation of Flynn to Trump's eventual firing of Comey, we decided that we needed to widen the case to include obstruction. We also didn't wrestle with the question of whether we could do both a criminal and a counterintelligence investigation, because the days of bifurcated criminal and counterintelligence investigations had long since passed into pre-9/11 history.

As I've already mentioned, investigations of national security crimes necessarily involve both criminal and intelligence aspects. They are two sides of the same coin. The FBI doesn't just investigate someone for the crime of spying as defined by the Espionage Act. As we build a criminal case, we also seek to understand what information the foreign nation — in this case, Russia — is trying to obtain. How was Russia's interaction with Trump advancing its strategic interests in Europe, in Syria, around the world? Who was Russia using — intelligence officers, contractors, and/or oligarchs? How were they communicating? Were they using business or other election assistance to leverage favorable action? And how could we get ahead of Russia to protect the nation?

Even the potentially obstructive act of firing Comey had implications for national security. Comey was leading the agency investigating links between Russia and the administration. His firing didn't just impede the investigation. Because the investigation related to understanding Russia's acts and intentions, Comey's firing harmed our ability to get to the bottom of what Russia was doing — in effect harming national security more broadly.

These were among the easier questions to answer; others were much harder. Beyond the thorny legal issues about the FBI's authority to open an investigation of the president we had wrestled with, I still harbored serious questions about what an investigation of Trump would actually entail and what it would or could achieve. The boilerplate wording from our operational guidelines was clear enough: our job was to investigate whether Trump was or had been, wittingly or unwittingly, involved in activities for or on behalf of the government of Russia that might constitute violations of law or threats to national security. But what did that translate to, practically and philosophically, when the subject was the president? Asking those questions made my heart hurt, as well as my head.

POWERS AND DUTIES

Only a week had passed since Comey's firing, but it felt like months. No relief came from the onslaught of investigative developments — or of related news and public information, as the media raced alongside us and various entities in the federal government sprang leaks that threatened at any moment to turn into a flood.

Shortly after noon on Tuesday, May 16, the same day we opened a case on Trump, the *New York Times* published a bombshell story that had been part of our obstruction debate for days but hadn't yet leaked to the public. The article reported that Trump had asked Comey to stop the investigation of Flynn in February and that Comey had jotted his recollections of that meeting into a memo — one of the seven memos that, on the night he was fired, I had carefully placed into the FBI case system for safekeeping and later retrieval.

How this information got leaked to the press I didn't yet know, but I knew it had to be by someone who was aware that Comey had written memos to record his conversations with Trump. I was frustrated because we had no warning, which required us to scramble to react to what was now a very public matter. If there was any silver lining, it was that public exposure might force some sort of resolution, most likely the appointment of an independent special counsel who would be insulated from political pressure.

Less than a month later, in a stunning turn of events, Comey would reveal that he had been behind the disclosure of the memo. He had provided it to his friend Daniel Richman, a law professor at Columbia, who in turn conveyed it to the press — just as Comey had hoped he would, anticipating that the memo's release would spark the appointment of a special counsel. I was dismayed that we had to hear about it from his testimony rather than directly from Comey or his attorneys, even if, in this case, the publication helped to advance a cause in which I believed.

In a twist of supreme irony, moreover, Comey's memos would eventually face scrutiny for the same reason that Clinton's email did in Midyear: that they might contain classified information. When I read all of them, that was one of my first and immediate concerns. In the U.S. government, it is the responsibility of the person drafting a document to label it classified if necessary. As I read them, I sensed that two or three unmarked memos might contain material that warranted a low level of classification. The danger was not that they might contain some piece of highly classified information, but rather that they might hold something more along the lines of Comey's recounting a comment from the president about a foreign country or a comparison between countries that might damage national security or cause a rift with an ally if it became public.

I hounded our attorneys about that concern. As a result, I found myself sometime later sitting in a conference room that adjoined Baker's office in headquarters. Somewhat poetically, it was called the Sunshine Room because of the warm light that spilled through the west-facing windows looking out on the interior courtyard. Baker and a small group of attorneys sat around the table with several copies of Comey's memos spread out in front of us.

As we walked through Comey's seven memos, I argued why I thought several should be considered classified. At times Baker pushed back, testing my arguments, but eventually we all came to the same conclusion: beyond the initial memo, which recounted Comey's pre-inauguration briefing to Trump in New York and which Comey himself had classified at the secret level, a second memo contained secret information, and two more were classified at the confidential level.

As we sat in the Sunshine Room, I couldn't help but feel a bite of mocking déjà vu about where we found ourselves. Like some sort of cyclical Greek tragedy, we were having to apply the same test to our previous director's memos that we had used for Clinton's email. As best we could tell, what had been Comey's intent? What was the volume and classification level of the material? What was the context in which Comey had written them?

All of us knew that classifying the memos would open Comey up to allegations that, by not classifying them initially, as he should have, he had mishandled classified information and should be prosecuted. But that didn't play a role in our decisions. We ultimately decided that we weren't in a position to appropriately make the decision, so we notified the inspector general that we believed several of Comey's memos contained classified information, setting in motion a two-year investigation that would end in August 2019 with a harsh condemnation of Comey for setting "a dangerous example for the over 35,000 current FBI employees." The utter lack of prosecutive merit — a view that attorney general William Barr's Department of Justice finally confirmed in 2019 — wouldn't matter to the political spin that followed the release of the IG's report.

The partisan noise over the memos drowned out another crucial irony: our scrutiny of Comey's memo was fundamentally incompatible with the notion of some deep-state conspiracy. If we had wanted to protect Comey, we could have argued that the memos weren't classified, or simply let the issue linger, unaddressed and unresolved. But that isn't how we ever approached our job. Rather, we asked, what do the law and regulations say? It didn't matter who did it, their rank, their political party, or their relationship to any of us. The only thing to do was the right thing to do.

But on May 16, 2017, that furor over Comey's handling of the memos was still far in the future. As news of Comey's memo burned up the Internet, Bill and I joined a small group in the FBI chief of staff's office to learn about another stunning development that had surfaced that very morning during a seemingly manic conversation between Rosenstein and McCabe. We listened to a recounting of a startling series of comments that the DAG had made to McCabe in a meeting earlier that day. In a now-famous exchange, Rosenstein had told McCabe that Trump had initially written a letter firing Comey, which White House counsel Don McGahn had advised the president not to send; this led to Trump's request for Rosenstein to write a letter instead, which Trump then used as a pretext for firing the FBI director. As the morning's meeting had proceeded and Rosenstein had jumped between topics with McCabe, another person in the room at DOJ — someone un-named in the meeting notes — said Trump was "not fit to be President," while another confessed, "I fear for the Republic."

We sat in silence, valiantly trying to preserve our poker faces as we heard about Rosenstein's offer to wear a wire at the White House to collect evidence of Trump's true intentions. At the time my sense — later confirmed by McCabe himself — was that McCabe took Rosenstein completely at his word, though DOJ later tried to spin the comments as dark humor from the DAG.

As if this weren't shocking enough, at the end of the conversation Rosenstein had shifted topics again. This time he brought up the 25th Amendment to the Constitution. The 25th Amendment is essentially the constitutional nuclear option for removing a president if he "is unable to discharge the powers and duties of his office." Rosenstein had done the math and knew that the Constitution would require 8 of 15 cabinet officials to agree. He had speculated that he might have two votes, in Sessions and then–secretary of homeland security General John Kelly, adding that Kelly had offered to resign to join in taking a stand against Trump.

As the recitation of Rosenstein's comments ended, we all sat very still. I didn't know what to make of Rosenstein, and still don't. At one point Comey described him as "a survivor." Rosenstein was clearly deeply concerned about Trump's actions, yet had affirmatively participated in enabling them. He was mistrustful of the Bureau's motivations, yet later appeared to keep the special counsel in place until he ended his work. Above all, in the many months to come he had multiple meetings with Trump during the height of the president's rage over the continued investigations, yet still emerged with his job intact and ultimately put himself and his reputation behind DOJ's deceitful announcement of the findings of the Mueller Report.

When Rosenstein's remarks to McCabe during this meeting in mid-2017 became known over a year later, in the fall of 2018, the resulting histrionics, particularly from Congress, included the suggestion that the discussion of the 25th Amendment amounted to a "coup." That assertion is ludicrous. First, as was clear to everyone who heard the meeting minutes on the day they occurred, this was not a detailed planning session so much as Rosenstein musing out loud, while under tremendous pressure, as he tried to sort out the best path forward.

A second obvious point belies the notion that this was a coup: it was about *the 25th Amendment of the U.S. Constitution.* This was not a whispered plot about imposing martial law or frog-marching Trump out of the White House at gunpoint; it was a discussion about a lawful means to address the scenario of a president whom the vice president and a majority of his cabinet felt could not discharge the powers and duties of the presidency. This emergency escape valve, if set in motion and contested by the president, would still require support from two-thirds of both houses of Congress. And the deputy attorney general of the United States — actually, the acting attorney general in this case, as Sessions had recused himself from all this — had just expressed (in what McCabe took to be a serious manner) his concern that Trump may not have been capable of carrying out his constitutional duties. In that event, other public officials had the ability and duty — under the Constitution — to remove him. This was the opposite of a coup, as it was a constitutionally contemplated mechanism for the peaceful transfer of power.

I sat in my office later that afternoon puzzling through all the many problems with Rosenstein's offer to wear a wire. It seemed wrong. What would he achieve by recording the president? Who would provide the operational authority to record? What field office would provide the equipment, and how in God's name would we brief its staff about what was going on without the news spreading like wildfire? I saw a lot of impossible questions, a lot of risk, and not a lot of benefit to the DAG's idea.

Fortunately, Baker independently arrived at the same conclusion and quickly advised McCabe that we weren't at the point where we should be considering Rosenstein's offer. To the best of my knowledge, the suggestion never came up again.

A BUDGET GUY

With investigations now open against both the president and the attorney general, all of us wanted to be assiduous in our approach, following not only the letter but the spirit of the rules and laws under which we operate. To my mind, that meant briefing Congress about these two new cases. Bipartisan support from the leadership of the legislative branch not only seemed appropriate but was needed to further establish the authority and the righteousness of what we were doing and to protect against unfair accusations that the FBI was somehow running wild.

So I suggested that McCabe, ideally accompanied by Rosenstein, brief the Gang of Eight, the group comprising the Speaker of the House, the House minority leader, the Senate majority and minority leaders, and the majority and minority leaders of the House and Senate Intelligence Committees. Others may well have had the same thought, and regardless, the suggestion was quickly accepted. As our Office of Congressional Affairs arranged a meeting time with the congressional leadership, we began circulating talking points with Rosenstein's staff outlining the scope of McCabe's briefing about the two cases. The Office of Congressional Affairs was able to schedule the meeting on the Hill for the next day, May 17. The briefing would include McCabe, Rosenstein, and their legal advisers and that was it; I was not among those invited, and would need to wait until they returned to hear the results.

With a congressional briefing mere hours away, we accelerated our preparation, passing drafts back and forth on email late into the night. DOJ adjusted McCabe's talking points, nominally to ensure that the logic of our each and every investigative decision was clear. We had no idea about an explosive topic they were about to add into the mix.

The following afternoon McCabe and a small team pulled out of FBI headquarters onto 10th Street in a convoy of black SUVs and turned left onto Pennsylvania Avenue for the short drive to Capitol Hill. En route, Rosenstein contacted McCabe with stunning news.

The DAG, who was in his own vehicle heading to the Hill, had called the White House moments earlier to report that he had just appointed former FBI director Robert S. Mueller III as special counsel for the Russia investigation. Rosenstein suggested that McGahn tell Trump and recommended that Trump make a statement welcoming the appointment. Needless to say, Trump did no such thing.

After passing through the Capitol Police security checkpoints and making their separate ways to the House basement, McCabe and then Rosenstein took their places in the secure briefing room, along with the Gang of Eight and their staff. We hadn't been expecting one of the eight, House Intelligence Committee chairman Devin Nunes, to attend the meeting, but he eventually lumbered into the room. The California Republican had been forced to recuse himself from matters relating to the Russian attacks on the elections because of an ethics investigation into his relationship with the White House. Now, asked to recuse himself from the briefing, Nunes refused. *Great,* I thought when I heard about Nunes. *The White House will know every detail McCabe briefed before the sun goes down.*

McCabe went first, leading the members through the cases we had opened on Trump and Sessions. He explained the allegations against each of them, the intelligence and information we had on them, and how that evidence met the legal standards for opening investigations into them. After McCabe finished, Rosenstein addressed the members and announced the appointment of Mueller to investigate the Trump and Sessions cases, along with others in the Russia probe, which Comey had briefed them about earlier that spring.

Shortly after the May 17 meeting, I learned that House Speaker Paul Ryan stated that he would defer to the Intelligence Committee heads about the intelligence aspects of the briefing. He said, *I'm a budget guy,* which I interpreted as, *Corruption of the president is not in my wheelhouse.* I was dispirited when I heard about that comment, because of the historical gravity of what had just been presented to him. *You're the Speaker of the House,* I thought. *You're second in the presidential line of succession. You were just told*

that the FBI opened a case on the president of the United States and that a spe-
cial counsel will be leading it. There is no sitting this one out for your country.
Either object to this investigation or stand up to support it.

Ryan's avoidance aside, the fact was that at the time, not a single member of the Gang of Eight, Republican or Democrat, raised concerns about our investigations. No one suggested that there had been insufficient reason to open the cases. No one questioned their constitutionality or legality. Critics later rushed to question the motives and propriety of the Bureau's opening the cases on Trump and Sessions, and eventually on the origins of the Russia investigation, as these all became the subjects of their own investigations. If I had had any say over it, I'd have suggested to these "investigators of the investigators" that their first two witnesses should be Senate majority leader Mitch McConnell and Speaker Ryan. They were briefed on what we were doing within hours of the decision to investigate Trump, within mere minutes of the appointment of the special counsel, and previously about the investigations of Page, Manafort, Papadopoulos, and Flynn. They were told why we felt we needed to do it, and they had all their questions answered. They expressed no concerns or reservations, nor did anyone else in the Gang of Eight. At the time they didn't seem to think that what we were doing was inappropriate. Only the eventual winds of politics changed that apparent consensus.

TOTALITY

With Congress notified about the Trump and Sessions investigations and the news of Mueller's appointment, McCabe and his FBI entourage climbed into their SUVs and drove back down the hill to headquarters, where work began immediately on the next big task: setting up the special counsel's office.

I wasn't sure at this point whether I would be included on Mueller's investigative team; it was far too early to know who he would pick to support him. But as one of the leaders of the investigation that Mueller was now taking over, I had no doubt that I would be working closely with him and his staff, at least until the special counsel's own investigation got off the ground. And Mueller's first staffing decisions gave me confidence that whether or not I found a place on his team, the investigation that I had helped to start would be in very good hands going forward.

I was elated to learn that Aaron Zebley, an attorney I had known since he was a federal prosecutor in northern Virginia, was going to be Mueller's deputy. Almost a decade earlier, Aaron and I had worked together when he was a prosecutor and I was a squad supervisor, bringing a series of espionage cases to a four-conviction conclusion. He had served as FBI chief of staff when Mueller was director and had followed him into private practice in Washington, D.C. Zebley was the perfect choice — a thoughtful, soft-spoken, and indefatigable attorney who was also one of the brightest people I had ever met inside or outside government. If anyone could help us navigate the legal and political waters and interplay with DOJ, all the while conducting a comprehensive and fair investigation, it was him.

The following morning, on Thursday, May 18, I was on the phone with Aaron, who explained that he and Mueller hoped to get up to speed on what we had been doing as quickly as possible. No one wanted to waste any time. An initial meeting with Mueller had been scheduled for that afternoon but was bumped to May 19, which gave me welcome time to review details of our almost 10 months of work and plan how best to convey it to the special counsel.

The delay also provided me with a chance to step back and review what we knew about the case in its entirety. I spent most days deep in the details of individual cases, deciding how to move the investigations forward. It was rare to get an opportunity to stop and survey the entirety of our casework. But even with the extra day, there was little time to think through a *tour d'horizon* that we would need to present to Mueller. We could only make the many different threads of our investigation intelligible by knitting them into a narrative of what the Russians were trying to achieve, along with all the investigative avenues we were pursuing.

In my office I turned on cable news and sat with my files open as I began to organize those pieces into a complete picture for Mueller. Even with my intimate knowledge of the case, I was flabbergasted anew by what I saw as I reviewed the cases in their entirety. The various Russian activities we had documented made up a dizzyingly complicated portrait of foreign interference, and I was momentarily overwhelmed by the task of putting it into a succinct and coherent storyline for the special counsel.

I called Derek, who would be briefing Mueller with me, and together we talked about how to structure our presentation, what to include and what to leave out. So many fragmented pieces arrayed themselves before us, including, just to name a few, the Russian hack and release of Democratic Party email; independent political actors and campaign affiliates back-channeling with WikiLeaks; a sustained and coordinated Russian attack on U.S. electoral systems; clandestine Russian exploitation of social media to target hot-button voter issues; a campaign foreign policy adviser who knew about the Russian hacks before we did and lied to us; another campaign adviser with longtime connections to Russian intelligence personnel who couldn't keep his story straight; a former campaign manager with huge debt and troubling ties to Russian and pro-Russian Ukrainian government figures; a former national security advisor who had lied to us about his contacts with Russia and had been forced to resign because of it; an attorney general who had had contacts with Russia that he had not disclosed during his confirmation hearings.

And on top of it all, at the pinnacle of this heap of perfidy and treachery, sat a president who had lied to the public, cozied up to Russia, and, once he became aware of them, attempted to block our investigations at every turn.

As I wrestled with the briefing outline, I looked up to the television as I heard the voice of the president. The TV was tuned to a press conference with Trump and Colombian president Juan Manuel Santos. A reporter asked, "A lot of people would like to get to the bottom of a couple of things, give you a chance to go on record here. Did you at any time urge former FBI director James Comey in any way, shape, or form to close or to back down the investigation into Michael Flynn?" Trump, contradicting Comey, stuck with his denial: "No. No. Next question."

Well, I thought, *either he's lying or Comey's lying.* I didn't believe Comey was lying. And Trump's troubled relationship with the truth was by that point well documented.

The coincidental timing of Trump's comment as I worked on the special counsel briefing was weirdly fitting. Trump's answer went to the heart of what could be the hardest question I faced the next day from Mueller: "What do you make of the totality of everything — is this a coordinated conspiracy?" The uncomfortable truth of the matter was that I didn't know. Despite everything that had occurred, we still didn't know what Trump knew, and the answer would likely come only from him or his inner circle. I was skeptical that all the different threads amounted to anything more than bumbling incompetence, a confederacy of dunces who were too dumb to collude, as someone joked. In my view they were most likely a collection of grifters pursuing individual personal interests: their own money- and power-driven agendas.

Yet I also believed, based on all that we had already uncovered, that Trump was compromised — badly and in a myriad of ways. He had a host of questionable business dealings and thus far had refused to release his financial records. He had had innumerable alleged extramarital affairs, and his personal attorney claimed to have paid hush money to some of the women involved to silence them. His Trump Foundation charity had illegally funneled millions of dollars to his campaign, which Trump was later ordered by a court to repay to other charities. (The foundation, meanwhile, was dissolved.)

Perhaps Trump's most egregious vulnerabilities stemmed from his lies about his Russian dealings. Putin knew he had lied. And Trump knew that Putin knew — a shared understanding that provided the framework for a potentially coercive relationship between the president of the United States and the leader of one of our greatest adversaries. This simple fact could explain something that made no sense otherwise: why Trump repeatedly did what he had just done at the press conference, choosing the course of action that made little sense in the context of U.S. security but that clearly benefited Russia.

Based on this pattern of facts, a counterintelligence agent would have to have been incompetent *not* to be worried that Russia might be exerting coercive influence over our new president. But that's our job. If there was one person in the country who could determine whether I was right to be concerned, it was Robert S. Mueller III.

14

The Special Counsel

J. EDGAR HOOVER was the longest-serving director of the FBI. Robert S. Mueller III was the second-longest, having entered the office a week before the September 11, 2001, attacks and remaining there until September 4, 2013. His twelve years as director of the FBI made him a towering figure in the Bureau's modern history, steering the agency through some of its most tumultuous moments and anchoring his legacy as a cornerstone of federal law enforcement. Within the Bureau, his leadership is viewed with something close to reverence among agents, both those who served under him and younger agents who came after. That's especially true among the senior leaders, many of whom rose through the Counterterrorism Division and remembered Mueller's relentless pace, exacting demand for detail, and unwillingness to suffer fools gladly. But while he had been the director of the FBI and had held a variety of senior positions at DOJ, he had never been a special counsel.

The history of the special counsel law is long and complicated, and I'll leave it to legal scholars to explain it in full. But there are important things to know about it. The most recent iteration, passed in 1999, has only two thresholds for the appointment. The first is that a special counsel can be appointed if there's a conflict of interest or some other "extraordinary circumstance" that prevents the DOJ from investigating or litigating a case. Between Sessions's recusal and Trump's firing of Comey — apparently over an investigation relating to an investigation of his administration, conducted by investigators and attorneys that he oversaw — we now had met that bar.

The other threshold for the appointment of a special counsel is that "it would be in the public interest" to have a special counsel take responsibility for the matter. The law also sets out a series of qualifications, including that he or she must be a lawyer, must be outside the government, and must have "a reputation for integrity and impartial decision making" and "appropriate experience to ensure . . . that the investigation will be conducted ably, expeditiously and thoroughly."

If there was anyone who met these criteria, it was Mueller. To the public, his background and experience are well known: an Ivy League graduate who served as a Marine in Vietnam, he has a straightforward character and honesty that are joined with an exacting demand for fact and little tolerance for nonsense, which have made him an object of admiration for some and of derision for others. For me, he was, and remains, the former.

Even before he was named special counsel in the Russia investigation, I viewed Mueller as the epitome of integrity and a pillar of institutionalism. He valued facts and he knew the law, and he approached both with intellectual rigor and an unflinching commitment to do what was right. He would want to know what we had done and where we were going next. There would be no emotion, no drama. There also would be no waiting period or wasted time. Mueller had gotten to work within hours of his appointment on May 17. We needed to get him up to speed as fast as we could.

As Derek and I prepared for the first briefing on May 19, McCabe and others seemed anxious. Not because we hadn't done an excellent job of preparing the investigative groundwork for him, but because of what I took to be their memory of how things could go wrong. *This won't be like briefing Comey,* was the unspoken message. I had heard stories of Mueller drilling unpredictably into minute details, catching unprepared agents off-guard. Whether Mueller did that out of a prosecutor's doggedness, to test the competence of the person briefing him, or out of an insatiable hunger for detail, I didn't know. But I wanted to make sure it didn't happen to me.

That afternoon seven of us filed into the director's conference room.

Mueller arrived with his deputy, Aaron Zebley, and his adviser Jim Quarles, who had until then been a senior partner in Mueller's law firm. The three of them were an impressive and daunting trio. Mueller is tall and lean, with a jaw like an anvil, a silver thatch of hair that he sweeps to the right, and fierce eyebrows over eyes that pick up everything around him. When he speaks, it is precise. There are no superfluous words; he does not extemporize. When I first met him after 9/11, he was already middle-aged but had the demeanor of a much younger man. Sixteen years later I could see the years etched onto his face, how his lines had deepened and some of his movements had slowed. But his piercing questions showed that he still had the same mental agility and intellect. As if to drive home that point, he dressed much as he had a decade earlier, in a white oxford shirt, dark suit, and conservative tie. I could think of no better embodiment of the word *gravitas*.

Like Zebley, Quarles brought a strong work ethic and razor-sharp intelligence to the special counsel's team. With his distinguished gray hair and wire-rimmed glasses, Quarles was a sort of legal consigliere to Mueller. He had a soft voice and an easy laugh, but he was tough — he had worked on the Watergate Special Prosecution Force decades earlier. Given that and his intervening legal experience, I quickly came to admire and respect his judgment and wisdom. As the work of the special counsel gained steam, Quarles took the lead on dealing with Trump's attorneys — a task that was usually aggravating and always tricky — as well as heading up the initial investigation into various aspects of obstruction.

Derek started the briefing, setting out the details of what we knew of Russia's actions and what our initial investigations had entailed. Then I took over, turning to Comey's memos and the investigation of Sessions. We laid out our understanding of Russia's strategic goals and how it had gone about achieving them in the 2016 elections, specifically concerning the Trump campaign. We detailed the Russians' use of social media, their intelligence agencies and nontraditional actors like contractors and oligarchs, and their multifaceted cyberattacks on both political parties and voting infrastructure. Then we discussed how we understood all the ways those forces had come into contact with the Trump campaign and administration. Between the two of us, we sketched out the highlights of what we knew about Russia and the election. It took more than two hours.

Derek and I were similar briefers, in that we didn't like providing a lot of handouts, because we would rather have the person being briefed listen to what we had to say rather than read along or, worse, read ahead. Our approach also allowed us to adjust the level of detail we were providing on the fly, based on the reaction we were getting from Mueller's team. It quickly became apparent that we needn't have worried about that in this briefing. Mueller's appetite for information was undiminished. Over the course of a couple of hours, he sat listening, taciturn, absorbing details of more than nine months of our labors. Zebley and Quarles flanked him, taking in every fact. Watching their quiet concentration, I knew from past experience that within weeks Aaron would probably have slept a sum total of seven hours and would know the facts of the individual investigations better than almost everyone at the FBI.

As we reached the end of the briefing, Mueller smiled at Derek and said, *I'd like your binder, please.* The good news was that Mueller was smiling. The bad news for Derek was that he wasn't leaving the room with his binder. I chuckled to myself. In the deskbound battles of FBI headquarters, good briefing binders, with double-checked facts and source documents, tabbed at key details, were guarded like gold. For the briefest of moments I watched Derek struggling internally — defy Mueller or relinquish the prized document? For any executive at DOJ, the White House, or Congress, Derek would have said he would make them a copy and keep his original. But this was Mueller. After the mildest of unsuccessful protest, Derek slid the binder across the table. *We'd appreciate whatever written material you can give us,* Mueller concluded. *Please make three copies of everything.* Everyone stood to leave.

And that was it. We were off and running. Seeing Mueller in place, his leadership group hungry and eager to absorb as much detail as fast as they could, was reassuring. Reinforcements — big, bad, heavily armored reinforcements — had arrived. I sensed that Derek felt the relief as well.

As we left I sidled up next to him, walking empty-handed out of the room. *Rookie mistake,* I offered.

Are you kidding? he shot back. *That was just for this briefing. I've got my* binder *binder back in my office.*

THE BEST AND THE BRIGHTEST

By early June, DOJ had identified temporary space for the special counsel on the ground floor of a nondescript government building two blocks east of FBI headquarters, hidden in plain sight on D Street NW across from a now-defunct Au Bon Pain, ensconced in a concrete slab building indistinguishable from all the others in Penn Quarter. An unobtrusive security desk monitored unremarkable occupants, and on the first floor, a long corridor led to doors related to sleepy government audit and administrative offices. At the end was a single windowless door with a bell to the side, watched by CCTV mounted at an angle and set into the ceiling, nearly invisible to an approaching visitor. The only thing to shatter the illusion of anonymity was the older mustachioed officer at the front entrance, who by the second week greeted all of us by our names.

The windowless space within consisted of mostly unoccupied cubicles in the open areas, with offices around the perimeter. At this early stage the suite wasn't staffed up, and the emptiness imparted a feeling of sterility. A few desks displayed children's notes scrawled in crayon and framed snapshots of spouses on vacation, but the workspace was mostly austere and devoid of family photos, home team sports trinkets, and the other kinds of warm personal clutter that adorn a long-term workspace. I claimed an empty cubicle where I could perch during my visits to the Special Counsel's Office, but my own office — and the remainder of my full-time job as the deputy assistant director of the Counterintelligence Division, which continued unbated — was still at FBI headquarters.

The ascetic feeling of the Penn Quarter suite wasn't an indication that there was some lack of interest in working for the special counsel or a shortage of people to do it. Just the opposite; the office was an empty vessel waiting to be filled. There was no lack of outstanding candidates who understood the importance of the special counsel's mission and were eager to occupy the desks and offices. During the early assembly of the team, Mueller, Quarles, and others regularly received calls from well-respected judicial and legal colleagues recommending top people in their fields. In some cases an applicant seeking a good reference had requested the call. Many others, though, were unsolicited and unprompted. References were calling out of a sense of obligation to ensure that the most highly qualified people filled out the special prosecutor's team to assist in uncovering what could be the gravest constitutional crisis in a generation. Lawyer by lawyer, investigator by investigator, Mueller plucked the brightest candidates to serve in his organization. It wasn't a single working group but rather independent ensembles of attorneys, agents, and analysts in teams focused on violations of law by various related subjects.

In addition to learning about all the cases we had opened in the course of our investigation so far, Mueller, Zebley, and Quarles spent much of the first few weeks sorting through the résumés of attorney applicants. While they relied on me and the FBI to identify the FBI personnel, they were focused on the lawyers. Candidates who rose to the top included numerous current or prior DOJ attorneys. Mueller would ask me to discreetly look into their reputations with the FBI agents with whom they had worked: How were they to work with? Were they particularly competent? What was their work ethic? Any worries or concerns? I flagged problematic candidates so that their flaws could be addressed during the selection process.

My task was to set up the group of dedicated FBI investigators who would be assigned to the Special Counsel's Office to conduct Mueller's investigative work and to act as the initial liaison between his office and the Bureau. The Special Counsel's Office would not be an independent, standalone entity. Just the opposite; it would be heavily reliant on DOJ and the FBI, which would provide the special counsel with investigators, administrative resources, and so on. Although I still hadn't been assigned to the special counsel—the senior person who would sit at the top of the FBI's personnel had not been selected, and I wouldn't get officially pulled in to lead the team for another few weeks — for the time being I was function-ing as the FBI team's de facto leader. I had the most leadership experience with the various investigations being subsumed into Mueller's work and was the logical person to create the FBI's staffing for the effort. The Bureau was filled with extraordinarily competent people; my main focus was on ensuring that we had the right balance of skill sets and that we bridged the criminal and counterintelligence aspects of the investigation.

The work of any special counsel is by definition focused on criminal be-havior. The special counsel regulations from 1999 focus on violations of law, and the scope of Mueller's appointment orders reflected that. The pros-ecutors whom he was bringing onto his team would lead the hard work of determining whether laws had been broken. It was my job to get the right FBI personnel to help with these criminal investigations, but I had a second goal as well: I wanted to make sure that counterintelligence aspects of the work would be sufficiently addressed. It wasn't enough to just "do coun-terintelligence" by simply sending what we might uncover about Russian intelligence activities back to the FBI to watch them from afar in a manner disconnected from Mueller's work. We needed people who understood in-telligence well enough to see that merely extracting intelligence from Muel-ler's efforts wasn't enough, that it was a two-way street — that we should be using intelligence activities to inform and advance our criminal cases as well. We needed the experience and smarts of FBI investigators who could cement the slipperiest aspects of CI work into cases bound for courtrooms. That joint experience was a precious commodity at the FBI. We needed to find a way to bring it in.

So I sat down with Bob, the supervisor who had been overseeing the day-to-day aspects of the Russian investigations for the past few months, including several investigations that would be taken over by Mueller. To-gether we sketched out the best FBI structure to aid the new office. The Bureau was going to provide investigative personnel to the special counsel, and it was up to us to decide what that would look like. To some extent we already knew where to look. The management structure of Crossfire had changed several times. Following the election, the investigators on Crossfire Hurricane had recognized that our sprint to find out as much as we could before November 8 had suddenly turned into a marathon. In response we had shifted the investigation's counterintelligence management structure to reflect that longer-scoped effort, moving oversight of much of the Russian counterintelligence work and the cases on Page and Manafort, among others, to the division's Russia branch. Responsibility for various aspects of the day-to-day investigation changed hands, the roles evolving as we sought the right balance between a more diffuse effort that took advantage of a wide range of expertise and a more centralized, coordinated effort. The latest permutation of the operational leadership structure was a more holistic effort being headed by Bob.

Bob and I had been Quantico classmates, but we'd had very different paths before and after the academy. I had come into counterintelligence from the military, while Bob had come into the Bureau from a national accounting firm and had worked on high-profile financial investigations in the FBI. We both felt strongly, however, that tracing money — even more than proving contacts with Russia — was likely to be the most critical investigative effort for Mueller's team. We knew about ample contacts between Trump's team and Russia, after all; what we didn't yet have was granular detail about those interactions, particularly the financial ones. Only by proving that Trump or his associates were not merely benefiting from but also actively soliciting or accepting things of value from the Russians could we begin to understand the dynamics between these parties — and begin to apprehend what Russia was up to and how we should respond.

Putting our heads together, Bob and I built a team representing a cross-section of criminal and counterintelligence expertise, buttressed by analysts with subject-matter expertise in areas like Russian intelligence services and oligarchs. To follow the money trail, we also set up a pool of forensic accountants. Good forensic accountants were and always had been in short supply in the FBI, because there was always more complex financial work than there were accountants — and because people with these sorts of accounting skills can make a lot more money in the private sector.

Talking with people we trusted, Bob and I identified about 60 FBI personnel who we thought would be a good fit for Mueller's team. Agents who had taken on everything from Wall Street to the SVR. Analysts who had spent time in Budapest tracking the transnational organized crime of Russian oligarchs. Accountants who could take registration documents of offshore shell companies and find the thread of the flow of illicit money. Professional staff, such as contract specialists and administrative assistants, who had worked on resource-intensive cases and knew the ins and outs of procurement, funding, and record-keeping.

We began making phone calls. No one was brought in without two independent recommendations. The selection took weeks and was an administrative headache. Often the people we contacted had commitments like a trial or the upcoming birth of a child that made the candidates hesitate, and led to the hardest question to answer, which was also one of the most relevant for people outside the D.C. area: *How long will this last?*

We answered as honestly as we could: no one knew. Depending on how the cases evolved, it could be several months or even years, if an investigation ended up going to trial. Despite those uncertainties, Bob and I were both gratified and relieved by how many FBI personnel agreed to answer the call, sacrificing priceless time with their families to come to D.C. and work for Mueller.

As Bob and I filled out the special counsel's team with seasoned FBI investigators, Aaron and I hammered out the details of the organization of the office. We placed lead attorneys on groups of related investigations, incorporating and augmenting the FBI personnel already assigned to those cases. Simply hiring attorneys and bringing over FBI personnel wasn't enough. We needed to create a structure that combined the right resources with the right investigations — dedicating some personnel to specific targets while creating shared groups, such as our forensic accountants, when resources were scarce. Most of all we needed a structure that would provide form and leadership to the entire effort. Rather than 20 ships navigating their own way toward a destination, we needed a focused and mutually supporting formation with the same objective.

One of the first attorneys Mueller brought on board was Andrew Weissmann, a DOJ attorney I had first met when he was the FBI's general counsel. Mueller hired Weissmann to lead the criminal investigations into Manafort and his associates, such as Rick Gates. (This case outgrowth was typical: the Special Counsel's Office developed many investigations beyond the initial Crossfire cases, which included — but were not limited to — the four initial cases on Papadopoulos, Page, Manafort, and Flynn. These new cases include many that are known as of this writing, such as the investigation of Michael Cohen, as well as some that to this day have not been disclosed.) Weissmann had extensive prosecution experience working organized crime and complex white-collar crime cases, such as leading the task force investigating the Enron scandal, and he provided new energy to the Manafort investigation. Although we had made inroads into understanding the counterintelligence aspects of Manafort's connections to Russia and the Ukraine before Mueller's appointment, I had been increasingly worried that the criminal aspects of the investigation were getting stuck in a bureaucratic morass. Now Weissmann, along with a seasoned lead agent who had worked with him on Enron, kicked the criminal aspects of the investigation into high gear.

Jeannie Rhee also came over from the law firm Mueller had just left and took the lead on what we called the "core Russia" areas of criminal activity. She focused on the interaction between the campaign and Russia, including people such as Papadopoulos, Page, and Trump's personal attorney Michael Cohen, who we now knew had been involved in the Trump Tower Moscow project that Trump appeared to have lied about so brazenly. With her experience as a DOJ prosecutor as well as high-level work in private practice, Rhee brought a constant smile and an inexhaustible supply of energy and aggressiveness, both of which were necessary to tackle the sprawling scope of her work. Two quiet but relentless lead agents joined her. One was an extremely experienced white-collar crime agent from Texas, the second an agent who had successfully investigated and prosecuted complex Chinese espionage investigations and who had — until I called him back — successfully escaped Washington, D.C., for the FBI's office in his hometown of Macon, Georgia.

Brandon Van Grack, a prosecutor from the National Security Division at DOJ, handled Flynn's prosecution and related criminal investigations, like that of Bijan Rafiekian, which came about as a result of investigation into Flynn's work with the government of Turkey. Van Grack was an excellent prosecutor, well liked at WFO and FBI headquarters for his expertise, aggressiveness, and easy teamwork. Flynn's case agent, a diehard Pittsburgh Steelers fan who also happened to have been the case agent for Edward Snowden's investigation, joined Van Grack. They frequently worked with Zainab Ahmad, a hard-charging prosecutor from the Southern District of New York, who worked on several aspects of the office's work that remain undisclosed.

Rounding out the initial teams, Quarles and a supervisor from the Criminal Division at FBI headquarters led the effort of looking into Trump's possible obstruction. Trump's attempts to thwart the Russia investigation, up to and including firing the FBI director, had not just provided the raison d'etre for the Special Counsel's Office; ironically, they had also become the subject of a new line of investigation by Mueller's team. Their focus was very much on the potentially criminal aspects of Trump's suspected obstruction.

The broader counterintelligence concerns about the president — which included the ways in which his suspected obstruction of the Russia probe might have been coerced by or intended to aid Russia — were investigated by multiple teams. Even at the time that I left Mueller's team, we were still looking for the right way to investigate those counterintelligence concerns, rather than have them fall into a secondary role to the obstruction investigation. No one used the word *obstruction*, though; we referred to it instead as simply *Team 600*, an oblique reference to the section of law outlining the special counsel regulations. But when anyone around the office heard that term, we all knew what was being discussed.

Mueller's leadership included me, Quarles, Zebley, and DOJ deputy solicitor general Michael Dreeben, whom I had never met before the appointment of the special counsel. Mueller humorously and deferentially called him "Professor Dreeben," delivered with a dry, wry smile. Dreeben had argued more than 100 cases in front of the U.S. Supreme Court and was widely admired as one of the country's foremost experts in criminal law.

If my praise seems both uniform and glowing, it's because it's merited. Every single one of these people was hand-selected, drawn from an over-abundance of qualified candidates — all of whom, by virtue of being hired as an agent or analyst or prosecutor, had already survived a highly selective hiring process. In the rare cases when less-well-suited people might have been assigned to matters, we found a way to reassign them to other tasks.

Within a month we moved into permanent office space at Patriots Plaza, a series of office buildings south of the National Mall, not too far from the National Air and Space Museum. By then the office had eased into a more regular routine. Agents and analysts were trickling in from across the nation, sniffing out the best deals for long-term hotel stays, and snatching up the best cubicles in their part of the floor. New attorneys arrived and quickly got up to speed in the various groups.

Leadership meetings began and ended each day, while Mueller, Quarles, Zebley, and I met weekly with the investigative groups and their supervisors. While the four of us knew everything the office was doing, by design the various teams were kept separate to compartmentalize the investigations. Despite that, all the team leaders needed to meet regularly to coordinate issues relevant to the entire office, such as issuing subpoenas, which might become public knowledge and tip off a particular investigative avenue we were pursuing. Soon enough the public would find out what we knew, but we wanted to be the ones to decide when.

PATRIOTS PLAZA

As spring of 2017 turned to summer, our floor in Patriots Plaza bustled with the special counsel's work. In my experience, complex cases have a natural ebb and flow — at any given time, some investigative avenues are active while others lie dormant. That wasn't the way it worked in Mueller's office. Every investigation charged forward at a blistering pace. The attorneys began issuing scores of grand jury subpoenas, and by July we were actively serving search warrants, including one on July 19 related to Michael Cohen's email. Besides his work on the Moscow project, Cohen had other connections to Russia and had exhibited some questionable behavior. We suspected that these ties might be relevant to our investigation, and those suspicions would be confirmed later, when Cohen eventually began cooperating with the special counsel.

The volume of investigative work was staggering. It included some 500 witness interviews, more than 2,800 subpoenas, almost 500 search warrants, over 200 orders for communication records, and nearly 50 pen registers (a device or program that records incoming and outgoing phone calls or email, but just the number or account, not the content of the communication). But just relating it doesn't do justice to the additional work that was required to make sense of all the information that these investigative efforts produced. Getting all that data was a challenge, to be sure, but the success of any good investigation lies in the grind of reviewing every piece of that information, making connections to find the truth — and doing so in a timely and intentional manner. If you're not disciplined, it's easy to fall behind, and if you're out doing a third interview before you've written up the first two, you're putting yourself in a bad spot that can suddenly become much worse when 20 megabytes of subpoena returns come in.

Complicating the investigations further, we weren't just investigating past events; new developments were unfolding in real time, forcing us to constantly update and expand the scope of investigations. On June 1, Putin announced that Russia had played no role in the elections — a claim that we knew was false but that Trump buttressed by adopting an almost daily mantra that he had engaged in "no collusion, no obstruction." As we began to interview witnesses, we confirmed that before Comey's firing, Trump had asked intelligence community chiefs to publicly deny that there were any links between him and Russia. Trump's behavior was so outlandish that the NSA's deputy director, a career intelligence official, said it was the most unusual thing he had experienced in 40 years of government service.

What's more, as we pushed forward with our investigative work, we had competition — plenty of it. There were at least three separate investigative efforts under way: ours, various congressional investigations (notably those of the Senate Intelligence Committee), and those of the ravenous media. All pursued overlapping paths, all sought the same people to interview, and to some extent all were fighting to find information first.

It was like a three-way race that we didn't always win. On July 8, I stood on a gray-carpeted aisle running the length of cubicles in the Special Counsel's Office, watching cable news break into programming to announce a *New York Times* report that Donald Trump Jr., along with Manafort and Jared Kushner, had met a group of Russians — among them Natalia Veselnitskaya, the Kremlin-connected lawyer — at Trump Tower in New York in June 2016. That alone was explosive, given that Russia was at the time waging a shadow war against the Democratic candidate on behalf of Trump Sr. The media report was the first we'd heard of the meeting, and we couldn't know how much more explosive this revelation would prove to be.

The *Times* had a head start, but we were determined to catch up and immediately set out to identify the participants at the meeting, send agents to interview them, and cut subpoenas for their phone records. While we made headway on the past facts of the meeting, unearthing new links between Veselnitskaya and the government of Russia, the media pressed the administration as Trump fumbled the response to the breaking news. Trump Jr. initially issued a statement detailing the background to the meeting in Trump Tower, claiming that it was about American adoptions of Russian children. But it wouldn't take long for the truth to come out — and when it did, we would find ourselves racing to keep up once again.

Astonishingly, just three days after the *Times* broke the news about the Trump Tower meeting, Trump Jr. released on Twitter a series of email messages that included a remarkable exchange with a man named Rob Goldstone, a former tabloid journalist. Goldstone had informed Trump Jr. that the "Crown prosecutor of Russia" had offered to "provide the Trump campaign with some official documents and information that would incriminate Hillary," cautioning that it was "very high level and sensitive information but is part of Russia and its government's support for Mr. Trump." Trump Jr.'s shocking response was one of the most memorable quotes of the investigation: "If it's what you say I love it, especially later in the summer."

I cannot overstate the effect that this string of revelations had on our investigation of Trump and his associates. The *Times* reporting and the frantic sequence of denials and admissions it triggered dramatically expanded our understanding of who on the Trump team was aware of, and receptive to, Russia's overtures. We had known that Papadopoulos was in that ignominious group; now we knew that Donald Trump's son, campaign manager, and son-in-law were in it too.

We watched the media absorb the Trump camp's denials that the president had been involved in drafting Donald Jr.'s statement claiming that the Trump Tower meeting had been about adoptions. And then slowly the White House twisted and turned toward the truth: Trump had helped edit Don Jr.'s statement and had pushed White House personnel not to disclose the email to the public. Whether or not Trump knew the real reason for the Trump Tower meeting at the time, he certainly seemed to know about it after the fact — and he seemed to have done his best to cover it up. Again I had the irrepressible thought: these were not the actions of an innocent man.

Immediately after this episode in early July, Trump switched from defense to offense: he pivoted to attack Mueller, something he had until that point restrained himself from doing. The president repeatedly tweeted that the special counsel investigation was a "witch hunt" and went on Fox News to question Mueller's friendship with Comey. Although we didn't know it at the time, Trump also pressured McGahn to fire Mueller, prompting a threat from McGahn to resign. It felt like Trump was scrambling to undermine the investigation at every turn, employing the same strategy — and some of the same tactics — he had used when he pressured, then fired Comey.

The constant attacks from the White House and its surrogates reached a fever pitch in mid-July. As the clamor rose, Mueller assembled the entire special counsel crew to reassure the staff and calm any nerves. Mueller had never addressed the entire group since everyone had moved into the space a month or so earlier. He just wasn't a pep-talk kind of guy; he preferred stoic individual resolve to rah-rah group rallies. When Mueller's executive assistant sent out a short-notice email about the meeting, I walked through all the cubes in the office, shepherding everyone to the area outside Mueller's corner office. As we waited, almost everyone suspected that the meeting had to do with Trump's criticism; some in the group probably wondered if Mueller was about to announce that he had been fired.

As Mueller emerged from his office, the entire office fell silent. Looking calmly out across the dozens of faces in front of him, he began by acknowledging that he, like everyone else, had heard the speculation that Trump was going to fire him. Mueller reassured everyone that he had no information that that was happening, and while he couldn't predict what might happen, that didn't matter. Everyone, he went on, needed to keep focused on doing the work at hand as best as we could, tuning out distractions and focusing on our mission. That was what we could control.

Mueller's calm was contagious, and for seemingly the hundredth time in the past year, I was struck by how unusual the unfolding events were and how essential good, steady leadership was to navigating those challenges. After Mueller returned to his office, we all went back to our work. One advantage of the relentless pace was that there was little time to engage in speculative fretting. We had more important things to do.

SUNRISE, SUNSET

The sun broke over the Potomac River just after 6 a.m. on July 26 as I sat on the shore of Old Town, Alexandria, watching a motorboat putter across the waves. As I waited, I sipped a cup of coffee tinged with cream but no sugar, which I had bought earlier, standing with uniformed military men and women in line at Starbucks before heading to their jobs at the Pentagon. Putting the coffee down next to me on the weathered wooden bench, I looked down at my watch and thought, *Any minute now.*

Several blocks to the north, right next to a waterfront park, FBI agents waited patiently in cars next to Manafort's upscale midrise condominium.

One of the agents had sworn out an affidavit for a search warrant at the federal courthouse in Alexandria the day before. We were on the verge of taking a step that would leave no doubt in the public's mind about the special counsel's direction and progress: searching Trump's former campaign manager's house.

A few minutes after 6 a.m., the search leader gave the signal. On cue, several sedan and SUV doors sprang open and about a dozen agents raced toward Manafort's residence. They pounded on the door, announcing, "FBI, search warrant."

After getting a call that the search was under way, I tossed the half-empty coffee cup into a trash can and walked back to the car. I was there only as a precautionary measure, in case something unexpected or violent occurred. Back when I was still green, a seasoned assistant special agent in charge told me, *If you rise up the ranks, you're going to be tempted to insert yourself into arrests and searches, because they're fun. Don't. Make sure they're prepared, then trust your agents and supervisors to do their job.*

I waited as the agents secured the condo and began looking for documents and financial records related to potential criminal activity by Trump's former campaign manager. A year and a half later, a federal judge in the Eastern District of Virginia would curiously describe Manafort as having led an "otherwise blameless life." While that statement might be legally true, Manafort's life seemed the furthest thing from blameless to those of us investigating him at the time. He had lobbied in the U.S. for shady foreign leaders like strongman Ferdinand Marcos of the Philippines, the corrupt Mobutu Sese Seko from Zaire, and of course Viktor Yanukovych of Ukraine, the Russian-backed oligarch forced to flee after a popular uprising in 2014. We had good reason to believe that our search would turn up evidence of crimes — especially Manafort's financial and business records, which we hoped would give us insight into how he moved money around the world through innumerable LLCs and bank accounts.

We had pulled the trigger on the search out of concern that a recent request for information from Congress might cause Manafort to destroy or otherwise disturb evidence. We wanted to get it before anything happened to it. The day before, Manafort had met with the Senate Intelligence Committee, and the Senate Judiciary Committee had subpoenaed him for questioning. As with the *New York Times* story about the Trump Tower meeting, the parallel congressional investigation forced us to move more quickly than we otherwise might have.

I wasn't sure when news of the search would hit the press — and while I had no doubt that Trump's counsel would find out soon, publicizing the event didn't seem to be in anyone's interest. As the morning progressed, I drove by the condo complex. Finding it quiet, I continued north onto Route 1 to follow the Potomac up into Washington. At a stoplight I sent a brief note to Mueller and Zebley: *Search under way. No issues.*

The search team carted boxes of records and computer equipment from Manafort's condo back to WFO, which, unlike the Special Counsel's Office, had an evidence control room where we logged evidence into the file and stored it. As we began entering the material seized from Manafort's condo into evidence, we were faced with a new question. Given the risk — albeit small — that news of the Manafort search would leak to the media, should we apprehend Papadopoulos as well, while we had the chance?

Not long before, we had received a tip that the former campaign adviser was headed back into the U.S. He had been out of the country for some time, but we didn't know why. We were concerned that he was trying to dodge us, in part because of what he had done after meeting with us along with his attorney in February.

During that meeting, Papadopoulos had told us again that he was willing to cooperate with the investigation. Shortly after, though, he shut down his Facebook account, which he had maintained for over 10 years, then immediately created a new one. This was suspicious. So we obtained the old Facebook information and dug into his deletions. We discovered that the old account contained information about contacts he had had with Russian-connected individuals during the campaign. His Facebook activities had left naked evidence that he had lied to us and was continuing to do so.

We knew that we had enough evidence to charge Papadopoulos with lying to us, and it seemed to several of us that there was little advantage in letting him continue down the path he was on. Arresting him would both bring the case to a conclusion and hopefully convince him that the prospect of jail time was worth ending the games and being completely truthful with us. Now that Papadopoulos was about to reenter the country, we'd finally get that chance.

We decided to bring him in. We didn't have time to swear out a warrant, so agents and attorneys on the Papadopoulos team began the procedures for a warrantless arrest. That entails taking someone into custody without the usual warrant, then going to court after the fact with evidence that we had probable cause. In criminal cases they're not unusual, but a warrantless arrest in a counterintelligence case would make many national security attorneys at DOJ blanche. Our lawyers weren't worried. As one set of agents and attorneys sat around a conference table drafting out a criminal complaint, another set started planning Papadopoulos's arrest.

The day after watching the sun come up over Washington, I sat in my car watching the sun set at Dulles International Airport in northern Virginia. The radio was reporting news that the Senate had passed legislation earlier in the day imposing sanctions on the Kremlin — a powerful rebuke to Trump — while giving Congress additional power to prevent Trump from easing these punitive measures. The legislative branch, at least, was taking the Russia threat seriously. With that nagging sense of uncertainty that is the stock-in-trade of counterintelligence, I asked myself over and over, *Why isn't the executive branch standing up to Russia too?*

At twilight Papadopoulos's flight from Munich touched down and taxied toward the gate. Agents waited for him near Customs. After he stepped off the airplane, they pulled him aside and arrested him. Since we knew he had a lawyer, we couldn't question him on the spot. Instead one of our attorneys called his: *We've arrested George. We'll make every effort for you to talk to him as soon as we can, and hopefully we can talk after that.*

The arresting agents spent the night in an exhausting courtroom odyssey, moving Papadopoulos between the federal court in Virginia, where he had been arrested, and the District of Columbia courthouse where we would swear out an affidavit in support of a criminal complaint the next day. When I arrived at Patriots Plaza early the next morning, I saw the frustration on the faces of the fatigued agents who had been driving Papadopoulos around town trying to get him entered into the judicial process. *I'm sorry for the runaround last night,* I told them. Tired but pleased, I took them around the corner to a restaurant to buy them breakfast.

My satisfaction soured later that day, when I went back to my office at the special counsel to find a note on my desk. It was a message about a missed call, with a name I didn't recognize but an office that I did: a DOJ inspector general investigator, and she wanted to talk to me. Curious and a little anxious — no call from the IG is ever a good call — I punched in the number on my phone.

The investigator got straight to the point. The IG's office had been reviewing the Clinton email investigation in response to demands by Republicans in Congress and had begun looking at our communications during the Midyear investigation, including any text messages we had sent or received on our Bureau-issued devices. *We need to interview you to determine whether or not some of your text messages indicate improper political bias,* the investigator said.

Well, that's easy, I thought. *Of course they don't.* It was a startling request in more ways than one. For one thing, Midyear Exam was a thing of the past; we were now well beyond the 2016 election, and my head was now immersed in Russia. For another thing, I knew that I had conducted myself with complete professionalism at each stage of our investigations. Sure, I had personal political beliefs, just like anyone else. But it was a point of pride for me, as for the vast majority of FBI agents, that as we worked each day to defend our country, we were guided by our oath of office rather than by personal ideology.

But I didn't say any of that as I held the phone to my ear.

Sure, I simply said, with little appreciation of the ferocious storm that was about to break.

15

The Storm

AS I ANTICIPATED the looming IG interview, I thought back to the texts I had sent during the Midyear investigation. I didn't remember many of them. But I did know that I had expressed personal opinions about Trump as well as other officials and public figures. Sometimes I had couched those opinions in snarky, crude terms. For instance, my view of Trump — his bigotry, lack of empathy and qualification and competence, and what I perceived as fealty to his own interests rather than the country's — was clear and disapproving in the texts I sent during the 2016 presidential campaign. I knew that much. But as I thought back to this time, it was also clear to me that my negative texts weren't limited to him.

Agents frequently don't like the people they investigate. What's more, I am interested in politics and national security, and my texts reflected that. I grumbled about the games that the Clinton camp played. I griped over just about everything that involved DOJ. While watching the statements coming out of the primaries and the presidential campaigns, I offered candid observations about Democrats and Republicans alike. Like almost every FBI agent, I had opinions about politics. And like every FBI agent I encountered during my career, those opinions were checked at the door when we walked into work in the morning.

But still — as I thought back on my messages, I felt a pit in my stomach. I knew they were written in a shorthand common for text messages that might be ripe for misinterpretation. And the more I dwelled on that thought, the more the pit grew.

Make no mistake—I regret sending those texts. But there's a point that's worth noting. Bureau policy allowed personal use of FBI smartphones, and most everyone I knew in the Bureau used their work phones to text or email friends and family, or even people like their doctor, clergy member, or attorney. So did I.

Having said that, just because something might be allowed doesn't necessarily make it wise or prudent. To be clear, I have made some terrible personal decisions, and they have hurt the people and institutions I love the most in the world—my wife, my family, and the Bureau—in a way that I'm deeply sorry for. At one point much later in the swirl of all this, my wife reflected to me, "You deserve to be divorced, not fired." She was absolutely right. I don't deserve the extraordinary support I have received. I'm beyond fortunate. To this day, as hard as it is to achieve, I am trying to live up to what I've always tried to do when I've done something wrong: acknowledge it, own it, and make it right.

But my family is innocent, and they deserve peace. For that reason, as I said in the introduction, this book isn't about my private life. As I have done throughout this ordeal, including in eventual congressional testimony, I am drawing a bright narrative line to separate my professional work from my personal life, to prevent blameless parties—especially my wife and children—from suffering further consequences from my actions.

But my texts have become central to the saga of the Russia investigation in a way that I could never have predicted and I certainly never intended. Looking ahead to the IG interview, I knew that office would go through every text I had written from or received on my FBI-issued devices during the investigation.

Initially I had no concerns about what the investigators would find as they looked at the texts and how we conducted Midyear. In fact, I wanted them to look hard, because I believed their findings would confirm what I knew—that the investigation had been thorough and exacting and complete. I was confident they would see our relentless aggressiveness as anything but biased in favor of Clinton. Furthermore, I knew that whatever I told the IG during my interview with the investigators would be backed up by everyone else they spoke to. And foolishly, I was certain that the IG's review would conclude that if some of the texts were ill-advised, they hadn't affected our approach to the investigation, our actions, or its outcome.

I was naive.

THE EYE

In the first week of August, after I'd gotten the call from the inspector general's investigator but well before my interview, Mueller called me into his office to talk about the IG's investigation. The afternoon sun filtered in through his office windows as I sat down on the sofa next to his coffee table, and he came around from behind his desk to sit by my side on a government-procured leather armchair. He didn't take his usual seat, and the meeting that followed was anything but normal.

I was already ashamed and embarrassed that my private communications had been dragged into a discussion among senior members of the special counsel team and FBI, although that mortification paled in comparison to what would unfold several months later. Much more immediately, it would be eclipsed by a deep sadness, as the leader I so respected delivered the most difficult order I'd received in over a year of working on the Russia investigation.

If Mueller was angry, he didn't show it. In a soft voice, he explained that he had spoken with the inspector general, Michael Horowitz. Based on the investigation of my texts and the need for the special counsel's team to avoid distraction, Mueller required me to leave the Special Counsel's Office and return to the FBI. I agreed.

The last thing Mueller needed was to have to defend the work of his office when the IG eventually released a report. We had made extraordinary efforts to minimize even a whiff of partisanship, which had already become the target of cynical, politically expedient claims. Trump and others were savaging attorneys on the special counsel staff who had so much as donated to or done prior work for Democrats (unsurprisingly, these critics raised no concern about donations to Republican candidates). I wasn't concerned that the IG would find I had done anything improper, but I could see that the existence of negative texts about Trump would be twisted and misused if I stayed. While I would have given anything to have continued that important work, I knew I had to leave.

Most of those thoughts went unsaid, and the meeting was remarkably low-key. No drama, no theatrics. Given the circumstances and Mueller's reputation as a taciturn leader, I was caught off-guard by his gentle dismissal, the decency it showed, and the way it reflected his character and leadership. In some ways it would have been easier if he had dressed me down and yelled at me. I left his office with an even greater respect for him, and an even sharper regret for letting him down.

Soon after I met with Mueller, I received another unwelcome message. This one was to set up a meeting early the following week with FBI acting deputy director Dave Bowdich. He still occupied his office from when he had been associate deputy director. Comey had been fired months ago, but no one on the seventh floor had moved offices. Down the hall, the director's office remained empty, waiting for the arrival of Chris Wray, who would be sworn in as FBI director later that week.

When I got to Bowdich's office in the first week of August, I took a seat on the couch. Sitting across from me, he explained that he had been given the unenviable task of figuring out where to place me back in the Bureau, now that I had been removed from the Special Counsel's Office. *I'm transferring you to Human Resources,* he told me.

I was stunned. I protested that my career had always been in investigative positions. My former Counterintelligence Division job had been filled after I left for the special counsel, so I couldn't return to that position. I argued to Bowdich that a job in an operational division would make more sense.

Not unsympathetically, Bowdich said that his current job was essentially administrative as well, and he encouraged me to dive into the new role. *You're going to be scathed by this,* he warned, *and some of that may splash on the Bureau. But this isn't forever.* The message was clear: the decision was made and final, and at some point the storm clouds would pass and I would return to operational work. I needed to salute, work hard, and keep my head down. So I did.

Although I didn't know it at the time, gears were turning behind the scenes of the FBI leadership because of another crisis. Around the same time the IG's office told Mueller about my texts, his staff also informed McCabe about them. During the conversation in which they delivered the news about my texts to the acting director, the IG brought up a *Wall Street Journal* article that reported on the FBI's investigation of the Clinton Foundation. McCabe, who later explained that he had been caught off-guard by the line of questioning, initially told the IG that he did not recall authorizing the release of the information. After thinking about it over the weekend, McCabe called the IG back to correct what he said, informing the IG that he did in fact recall authorizing the release of the information after having had time to think. Much later, that interaction with the IG would be used as one justification for firing McCabe hours before he was eligible to retire.

Throughout the summer of 2017, following Comey's firing, Trump had grown increasingly hostile toward McCabe. The president privately questioned his loyalty while publicly tweeting for Sessions to replace him. Maintaining the drumbeat of pressure, Sessions asked Wray, shortly after he was confirmed as the Bureau's new director, to fire McCabe. Wray refused, telling Sessions that he would resign rather than be pressured into making politically motivated personnel decisions. It was not until I learned of all these details from the IG's report on McCabe in April 2018, and from McCabe's subsequent 2019 lawsuit disputing his firing, that I realized the scope of the political and organizational machinations that were going on behind the scenes at the Bureau. In mid-2017, not only was I unaware of that internal drama; I had yet to fully appreciate the way fear of Trump was throwing sand into the gears, preventing the Bureau from properly protecting its people.

During the fall I watched the special counsel's work from the outside, sidelined and frustrated as the investigations continued without my participation and beyond my ken. I often found myself scrolling through news online as the scope of Russia's social media manipulation began to become apparent. Russia hadn't just relied on troll farms to amplify stories. It had engaged in a long-term, complex effort to pit Americans against one another and drive wedges through the country, and we had been largely unaware of it. It wasn't just the FBI, either. The efforts of our entire intelligence community had come up short in that regard, and social media companies hadn't been vigilant or had failed to reveal the findings of internal reviews of their platforms. In particular I was irate that Facebook had taken as long as it had to share internal information about the Russian efforts and aggravated that we hadn't grasped it earlier. Most of all, I was angry and disappointed that I couldn't help fix the problem.

I immersed myself in my new Human Resources position. The best part of the job was the many talented and dedicated people, who were doing complicated work. For example, behind the scenes they were headhunting and recruiting new agents and analysts with skill sets that the Bureau would need not only next year but in the next decade and beyond to address, among other things, the cyber threat that we had missed. They were also working to increase diversity and rethinking what "good" performance was and how to measure it. They were planning for a future long after Crossfire Hurricane was a thing of the past.

Then an actual storm hit. That fall Hurricane Maria barreled over Puerto Rico and the U.S. Virgin Islands, devastating the islands and their inhabitants, including our personnel and offices. Immediately afterward, the San Juan airport closed to commercial air traffic; basic services such as power, water, and communications were disrupted on an immense scale. I was allowed to lead the Bureau's Human Resources response to the disaster, coordinating the relocation of uprooted families and finding and deploying Bureau personnel to continue the work of the field office as the hurricane recovery began.

At the beginning of October, I boarded a Bureau jet with Bowdich and a small team to fly down to Puerto Rico to survey the damage and assess our response. The island had been wiped out. Even though Maria had slammed into it almost a week and a half earlier, roads to many parts of the interior were still cut off. FBI helicopters helped by taking food and water to inaccessible areas, while agents worked on the ground with chainsaws to clear roads, all the while providing security where needed. We went by helicopter to our satellite offices on St. Croix and St. Thomas, both of which, like the surrounding islands, had been hit hard. At our tiny office in St. Croix, an agent who had lost the roof of his house in the storm refused the Bureau's offer to relocate to safer ground, instead moving his family into a sheltered corner of their home while they waited for repair crews. His wife was a midwife, and as he pointed out, life hadn't stopped: the hurricane hadn't delayed the arrival of babies. So there they were, in the midst of the debris and destruction, with no power, no telephone or cellular service and only intermittent running water. I was witnessing another example of how an agent and his family were diving in to help their community.

At the end of the trip, our convoy snaked through San Juan traffic, heading from our downtown office toward the airport for our flight back to D.C. Near the airport, the traffic slowed. Parked cars choked the shoulder of the highway, their occupants sitting on truck flatbeds or on folding chairs. It took me half a beat to figure out what was going on. Then I realized: people were waiting to catch a glimpse of Air Force One, which was bringing Trump for a visit to the storm-shattered island.

We managed to get into the private side of the airport before he arrived, and we settled into our FBI aircraft to wait for clearance to depart once the airspace was reopened. Other than the West Wing encounter during Flynn's interview, it was the closest I'd ever come to Trump, sitting on a plane that shared the same tarmac. He deplaned without my seeing him and his visit got under way as we taxied to the runway.

Our plane's engines surged to life; we accelerated down the runway and lifted off. Below us, at a San Juan church used to distribute disaster relief supplies, Trump spoke briefly, then — inexplicably, given the environment — made jokes and shot rolls of paper towels into the crowd as if they were bundled-up T-shirts at a basketball game.

As our jet climbed out over the Atlantic, Bowdich leaned toward me and reassured me that moving me to Human Resources had been the right call.

It could have been worse, he said. *Just imagine if Trump had gotten ahold of your texts.* He pantomimed tweeting into a phone.

It felt like we were all whistling past a graveyard. We could only hope that Trump's restraint, deliberate or unwitting, would continue — and that although he might be tempted to interfere yet again in the investigations into him, our commander in chief would abide by both common sense and the unwritten rules about acceptable presidential behavior that had guided all his predecessors. Yet deep down we knew that he probably wouldn't.

Meanwhile, Mueller's investigation showed no signs of slowing down. Two days after I returned from Puerto Rico, Papadopoulos signed a plea agreement, admitting that he had lied to us about the extent, timing, and nature of contacts he had had with Russia while a member of the Trump campaign. He also confessed to lying about Mifsud, his relationship with a Russian woman he believed to have connections to Putin, and his contacts with Ivan Timofeev, a Russian with connections to the Russian Ministry of Foreign Affairs. At the end of October, Manafort and his business partner Rick Gates were hit with a 12-count indictment that included money laundering and making false statements. Finally, on December 1, Flynn pled guilty to lying to us 11 months earlier in his West Wing office.

I felt vindicated to have conclusions to the cases we had opened in August 2016. In trying to determine whether a single allegation was true, we had instead uncovered a host of criminal activity involving Trump's campaign advisers, all related to Russia. And it wasn't simply the volume, it was who they were. Trump's campaign manager. Trump's national security advisor. Trump's foreign policy adviser. I had never experienced such a high volume of successful prosecutions — or the speed of those prosecutions — in any other series of investigations in my career. It confirmed that we had been right to have taken the threat as gravely as we had.

Ironically, I remember thinking that the steady release of these facts actually helped Trump. Unlike with Clinton, whose campaign was dogged by a constant slow drip of the leaking of stolen email, with the drops interspersed with periods of calm, the revelations about Trump were delivered in an unending torrent in which new revelations not only mixed with older ones but also seemed to dilute their potency. Both were drawn-out processes, but the same elongation yielded vastly different results for the different politicians. The spread-out attention seemed — inaccurately — to heighten the severity of Clinton's behavior, while in Trump's case it actually served to diminish the outrageousness of his wrongdoing and that of his associates. We all appeared to be getting numb from the volume and sever-ity of the conduct. A new, disappointing, and dangerous norm was being created.

FAME

In early December 2017, the FBI's Office of Public Affairs called to ask me to stop by. Walking down the hallways near its office on the seventh floor, just around the corner from the director's suite, I glanced at the framed movie posters about the FBI spanning decades. I had come to respect the assistant head of the office, Mike Kortan, after watching the Office of Public Affairs in action with Midyear and working with him on other cases. He was an effective agent who had been in the job forever and whose fat Rolodex of Washington media contacts was accompanied by an inscrutable poker face.

By then I had long since spoken with the inspector general's investigators and had grown accustomed to waiting as the IG continued his inquiry. I still felt confident that his office wouldn't find any evidence of serious or intentional wrongdoing and that I would eventually be transferred back to an investigative role at the Bureau if I continued to keep my head down and worked hard at my new post.

Suffice it to say that I wasn't at all prepared for the conversation that unfolded when I reached Mike's office. Mincing no words, he conveyed dis-concerting news: the New York Times was asking about my text messages.

An angry knot twisted in my stomach. So far knowledge of the texts had been limited, as far as I knew, to a tiny number of people at the FBI, DOJ, and the IG's office — a smaller group, in fact, than the number aware of our

Russia investigations. Someone inside that group had obviously leaked the information to the *Times,* and I knew the clock was ticking before the story broke.

Who could have done this — and who *would* have done it? I didn't think anyone in the FBI, IG, or Special Counsel's Office had a motive to leak my texts, or even the fact of them, to the media. I saw no benefit to any of those groups from doing so; in fact, it would only have made their jobs harder.

But I knew that the IG had been providing updates about the investigation to DOJ leadership. As I thought about it in retrospect, the likely source behind the leak of my texts was someone at a political level — either an appointed official like the attorney general or deputy attorney general or one of their staff — or someone to whom they had spoken at the White House. My texts were simply too enticing as a partisan bludgeon; they would be too tempting for anyone looking to discredit the investigations surrounding Trump. So someone with those motives had weaponized them.

By the end of the week it was clear that four months of tension-laden silence were about to end. Unhappy that the *Times* had yet to publish the story, the leaker had also apparently shared the information with the *Washington Post.* The two outlets were now racing each other to publish stories, and ultimately released their articles within hours of each other on the first Saturday of December. "Mueller Removed Top Agent in Russia Inquiry Over Possible Anti-Trump Texts," read the *Times* headline when the story appeared. "Top FBI official assigned to Mueller's Russia probe said to have been removed after sending anti-Trump texts," screamed the *Post.*

Now that the news had finally broken, it spread like wildfire, turning the remainder of that weekend into a panicked blur. Anticipating a crush of the press, I moved one of our cars away from our house to a place we could get to it without being seen. A black pickup truck carrying a photographer pulled up in front of the house, quickly followed by a silver SUV and then others. I watched them adjusting their equipment, occasionally emerging from the vehicles to stretch their legs as they waited to get a picture of me, my wife, or my children.

With the attention also came threats. Random callers yelling about a deep-state plot. Twitter messages looking forward to my daily raping at Guantanamo Bay. Anonymous letters and, incongruously, Christmas cards telling me that I was under their senders' constant surveillance.

My family and I were on our own. Any hope for respect for our privacy or sense of security was gone. We were forced out of our home, lucky to slip away unseen as our neighborhood continued to fill with reporters, paparazzi, and who knows who else.

Waking up in an unfamiliar bed on Sunday, December 3, I experienced a surreal shock, which would soon become an absurdly all-too-common occurrence. As I shook off the cobwebs of sleep, casting around our home-away-from-home for the coffeemaker, I grabbed my cell phone. Looking down at the screen, I saw the green notification bubble of an incoming text from a friend. I typed in my password as the gurgle of the percolating water began passing through the coffee grounds. The text was brief: *Hey, you know the president just tweeted about you, right?*

My heart sank. I pulled up Trump's Twitter feed, and there it was. Starting at 6:45 that morning, three tweets complaining about imagined affronts and wrongdoings and me. These tweets were the first of hundreds of attacks that the president of the United States would launch against me over the coming months — a smear campaign that is ongoing at the time of this writing, and that seems clearly intended to distract from the controversies that continue to plague Trump's presidency.

I sat there staring at my name peppering the Twitter feed of the president, typed again and again by the man holding the most powerful office in the world. Up to this point I had become as dulled and desensitized as anyone to each new presidential outburst over Twitter, but it has a quality all of its own when it's about *you*. Subsequently, when people dismissively told me that the president was just having a tantrum, just acting like a child on a playground, it offered me no consolation. Indeed, it missed the point: this was the *president* of the United States of America. It was surreal and dangerous.

The coffee machine let out a final sigh of steam and fell silent, unnoticed. My head spun and I felt lost. What saved me was my wife telling me to put down my phone, which I did. At the breakfast table, I processed and talked about the tweets. I was reminded of what mattered most: family and doing the right thing. Focusing on all the good around us even in that moment, on the truth, and on the good work we had done at the FBI helped me to reorient myself even in the strange surroundings of our exile.

I immediately began to receive notes and calls from longtime colleagues, some of whom I hadn't heard from in years. The kindness extended to me in a moment of crisis was overwhelming, matched only by the humbling surprise I felt after getting messages of support from people whom I respected but didn't know well. Carl, my former colleague from Boston, told me to buck up as only he could, in his characteristically salty style: *Keep your chin up, your head down, and your mouth shut. These people aren't worthy of carrying your jock strap.*

We stayed away from our house for almost a week. The media crush remained on the street and sidewalk in front until the reporters grew tired of waiting and the photographers left without pictures and video. That week also brought phone calls, letters, and genuine Christmas cards from strangers. Most were positive, and I have a box in the study filled with kind notes from people across America. But a few were alarming.

As Trump's tweeting about me continued and as media coverage of the texts unfolded, a group from the FBI's Security Division began monitoring threats against me and my family. Most were innocuous, but at least one was serious enough that a group of three personal security experts called me, then stopped by my office to detail some commonsense safety precautions. Parts of the briefing brought back memories of growing up overseas and of later deployments around the world. I had always assumed that such safeguards would be necessary only outside the United States. I was wrong.

I listened as the security experts ticked through best practices. Know your surroundings. Learn and keep track of vehicles that belong to the neighborhood. Keep a log of the descriptions and license plates of those that don't. Walk around your car before getting in, and look in the wheel wells and at the underbody. Drive at random times on random routes in and out of the neighborhood. Be wary of unexpected packages. Watch for mail that's oddly wrapped or carries too much postage. Then, like a punch in the stomach, they suggested sharing all this information with my family.

I cannot tell you how wrenching it was to receive that advice; I certainly would not wish the feeling on anyone. I also cannot conceive of a sadder testament to Trump's America — fringe elements of a nation governed by violent invective and vindictive fury. The president's vitriol posed a physical threat to his targets, who now included me, my children, and my wife.

The nightmare hasn't ended. For a long time, whenever we got ready to leave the house or heard a noise, we would peek out through the closed shades, hoping that no one was waiting outside. Strangers, among them photographers and "journalists," have followed, emailed, and called us at home and at work. After swearing to neighbors that they only wanted a picture of me and would leave "the wife and the girl" alone, some of these people chased my daughter and my wife to their car, cameras flashing. My family, driving past a police officer who could do nothing, had to back down the street to escape, because they didn't feel safe enough to stop to turn around.

The stress over Midyear Exam and Crossfire Hurricane used to wake me up before dawn. Now it was concern for my family's safety that kept me awake in the darkness. And as much as it sickened me to think it, my own government was partly to blame.

BALANCE OF POWERS

On January 11, 2018, I watched the floor numbers tick upward as a DOJ headquarters elevator carried me smoothly to an upper floor with my attorney, Aitan Goelman. It was just a few weeks after word about my texts had leaked, and I had already come to suspect that someone at DOJ was directly or indirectly behind it. Who, I had no idea — which made it more than a little unnerving to be venturing into the department's headquarters.

Stepping out of the elevator, I turned to look down the corridor — and felt a jolt as I spotted Rosenstein walking down the hall in my direction.

As we walked toward each other, I cast about for what to say to him, but there was no need. Before we passed, he stepped into a carpeted office and disappeared.

I continued down the hall, scanning the wall plaques next to each door for the room I was searching for. Almost everything about the DOJ building is *nicer* than FBI headquarters. Higher ceilings, fine art, large offices, smoked glass on old wooden doors, and floors that aren't linoleum. I felt a million miles away from the Hoover Building, in more ways than one.

I found the door I was looking for and opened it. Inside the quiet, wood-paneled conference room was a long wooden conference table with stacks of hundreds upon hundreds of pages of all types of records, waiting to be turned over to Congress. Among them, stacked up in page after page of printed paper, were my texts.

The reporting about my texts hadn't only whipped Trump into a frenzy. It had also sent Republicans in Congress into a righteous peeve, giving them fodder for right-wing indignation that would eventually ferment into the deep-state fairy tale that would consume conservative media. In December the immediate reaction of the Republican majority — ironically, the party I grew up with and had frequently voted for, as many, if not most, in law enforcement and national security do — reacted to the news by demanding copies of the texts. DOJ seemed all too happy to accommodate them. Just eleven days had passed since the news had broken about my removal from the Russia probe and already my superiors in the executive branch hierarchy were leading me toward the slaughterhouse.

Adding insult to injury, and in clear violation of law and precedent, DOJ personnel rushed to get an initial selection of texts out to the media in the middle of the night before a congressional appearance by Rosenstein on December 13, in an apparent attempt to give him cover from criticism Trump might lob his way as a result of the contentious testimony. What's more, they released this sampling of my texts to reporters with an underhanded insistence that the press could not attribute that access to DOJ. This fit a disturbing pattern, and only reinforced my conviction that elements of the department were behind the initial leaks about me to the media. Clearly those elements were happy to leak information when it suited them and then to avoid investigating or naming the leaker. Making matters worse, to this day no one has investigated the leaks about my texts to the *Times* and the *Post* (or, for that matter, resolved the leaks to Giuliani about the Weiner laptop during Midyear). Political expediency, not justice, was beginning to seem like the top priority for DOJ.

Beginning with these initial releases in mid-December, DOJ began a rolling production of records that Hill Republicans demanded, including some types of material the FBI had never turned over to Congress. Talking one morning to a senior attorney in the Office of General Counsel, I began to understand the scope of information that was going to be turned over. Not just my texts, but those of others — supervisors, agents, analysts. Email. Case files. Intelligence from foreign partners. Information that could be used to identify assets. Shaking my head, I bitterly noted that Rosenstein was leading DOJ away from decades of important and well-thought-out past practice of not releasing detailed records such as these to Congress just months after excoriating Comey for breaking with precedent. A person's consistency — or lack thereof — in applying principles says a great deal about that person's character. There's a fine line between nobility and hypocrisy when you're the one facing a 500-year flood.

I was outraged about all of this, but especially furious at DOJ's decision to release my texts amid an ongoing IG investigation. I understand how someone could look at the texts and wonder whether or not my personal beliefs affected my actions, but that's the issue the IG was tasked with and was actively addressing. DOJ tossed the texts into the middle of a partisan feeding frenzy, knowing that my hands were tied, as I was unable to address or defend my actions as an FBI agent while I waited another half a year for the IG's conclusions. I can't imagine that DOJ didn't know it was creating a deeply biased and politically charged environment for the IG's office to complete its investigation. Though a congressional subpoena might have forced DOJ's hand, based on prior and future practice, DOJ typically would have argued that the texts be withheld while investigations were pending.

In my opinion, the IG's office did not acquit itself well on this front either. When asked by DOJ for its position on releasing the texts, the IG had responded, in what I felt was a politically expedient and atypical way, that it had no objection to DOJ providing the never-before-seen material to Congress. Normally scrupulously tight-lipped during an investigation to avoid prejudice and protect its ability to investigate, the IG had scores of interviews to conduct and months of extensive investigation remaining. There was no way the subsequent political weaponizing of the texts by Trump and others would not skew what otherwise could have been a neutral investigative playing field. But Rosenstein was headed to the Hill, and he needed some cover.

The release of the IG report was still some six months in the future at this point, but the distance did little to minimize the effect of this irregular and politicized decision to release the texts. By the time the IG got around to issuing its report that June, conservative media and political figures were loudly highlighting outrageously inaccurate interpretations of the texts at every turn, twisting my phrases and taking my comments, even my humor, out of context to score political points. Senator Ron Johnson touted ridiculous allegations of "offsite meetings" of a "secret society," while House Republicans picked up on the absurd idea of a coup hatched against the campaign more than a year before we actually began our investigations. Having spent six months whipping unfounded allegations of bias into a frenzy, Trump and his enablers had successfully destroyed one of the fundamental precepts of professional civil service: that people can hold personal political opinions and separate those opinions from their work. By the time the IG's office published its findings — that there was no evidence that any investigative act was influenced by bias — it made little difference. The damage had been done.

After the initial unusual releases of the texts and other investigative materials, beginning in December 2017, DOJ realized that it had failed to strip off some identifying personal details and names of private individuals, including me. So its lawyers agreed to our request to review further material before it was provided to Congress, which was what took me to the DOJ on January 11, 2018, with my attorney.

As a young prosecutor, Aitan Goelman had cut his teeth on the Oklahoma City bombing prosecutions. In his subsequent career, he moved between government service and private practice. When I was looking for an attorney, a friend who had supervised the squad next to mine at WFO reminded me about some of his interactions with Aitan. *That guy was a pain in the* ass! he said. It was a grudging, backhanded compliment — and it was the endorsement I needed. Having spent a career on the opposite side of people represented by high-powered counsel, I was just beginning to appreciate how critical highly competent personal legal advice was.

Following some preliminary discussion with a DOJ attorney, I got to work reading the documents. It was stressful. Having watched weeks of hyperbolic spin and outright lies, I knew how some of the material would be taken out of context and used to hurt the FBI and the special counsel. As the hours slipped by, my attention wandered. I remembered some of the texts, but many others I had no memory of sending or receiving. It was hard to stay focused for an extended period of time on such an extremely detailed task — ensuring that my name and any identifying personal details didn't make it into the texts that DOJ released to Congress — as I was distracted by the knowledge that I was reviewing page after page of ammunition that soon would be deployed against me, Mueller, and the Bureau.

After several hours we broke for a quick lunch at one of the restaurants peppering Penn Quarter. After lunch I returned to the piles, dismayed by our slow progress. At some point Aitan's phone rang. He answered. I didn't pay much attention as he listened. I glanced up at him as he finished the call.

In a measured tone that conveyed the incredible-is-the-new-normal feeling of the first year of the administration, he looked me in the eye. *In an interview with the* Wall Street Journal, *Trump just accused you of treason,* he said. *The* Journal *wants to know if we have any comment.*

We didn't speak for a moment. I was shocked, then livid, then scared. I had spent my career in the army and then the FBI fighting for the U.S., putting on a gun and a badge each day to protect and defend the Constitution. I'd done that while going out of my way to keep a low public profile, the better to do my work and serve my country. And now the leader of our nation — the man who was my president too — had publicly accused me of one of the few crimes enumerated in the Constitution, a crime punishable by death.

In his declaration that I had committed treason, Trump appeared to be relying on my use of the insurance policy analogy in a text message from August 15, 2016, shortly after we had opened Crossfire. As I explained in Chapter 7, we had a vigorous and ongoing debate within the FBI during the summer and fall of 2016 about the competing goals of protecting intelligence sources and methods and resolving the allegations of Russian offers of assistance to the Trump campaign. My position was that the FBI shouldn't base its investigative decisions on the likelihood of anyone's candidacy being successful or unsuccessful. If Trump was elected, we might face the grave possibility that one or more persons in illicit contact with the government of Russia would be named to national security positions within the new administration. We needed to follow the normal path, doing what the FBI does. We needed to investigate — not because we thought someone was guilty or innocent but because our job was to find the facts. And we needed to do this in a timely fashion commensurate with the risk that the subjects of our investigations could end up in national security positions in the administration.

The FBI and DOJ knew that my insurance policy text was part of a discussion about how to balance source protection against aggressive investigation. But they chose to stay silent about it, allowing Trump to seize on the false spin of conservative media and certain members of Congress — and now to throw around accusations of treason.

As an FBI employee, I was prohibited from commenting to the media, even on nonsense. So ordinarily my answer to such a request from any publication would have been no. Ordinarily a career spent in counterintelligence requires a stoic "No comment." Ordinarily that commitment to public silence pervades every level of our world, from the individual employee to the FBI to the intelligence community as a whole.

Every agent is accustomed to the fact that commentators may take issue with the conduct or outcome of his or her investigations. But it's insidiously different when it's coming from the president of the United States.

I didn't know how to respond. As I was about to find out, the FBI didn't either. The corrosiveness of Trump's presidency was not just the administration's willful violation of accepted past behavior but the resultant chaos it created throughout the U.S. government. Precedent — important at any time but particularly critical in times of crisis — no longer provided relevant guidance. As a result, most people did the most logical thing: they took the path of least personal and organizational risk and resistance.

This wasn't the first time that Trump had pushed the FBI into such a corner. In March 2017 a Trump tweetstorm had culminated with the question "How low has President Obama gone to tapp [sic] my phones during the very sacred election process." The assertion was completely false — there was no wiretap of Trump's phone or any such surveillance of the candidate for president of the United States. Like his subsequent accusations of me, it was an outrageous fabrication, and later that morning I had called Bill Priestap to discuss a response to Trump's false claim about President Obama. We agreed that it was wrong and that the assertion was immensely damaging to the trust of the American public in the FBI. Bill forwarded our confirmation up the chain, and Comey asked DOJ to issue a statement rebutting the claim. It didn't. It seemed like an easy decision, but six months would pass before a court filing by DOJ finally settled that there were no records of any wiretaps described in the tweet.

Remembering that episode, and still stunned and stinging from Trump's deadly serious charge against me, I stepped out of the review room into a vestibule next to the elevator bank to call Bill.

Hi, Bill, I'm sorry to bother you, I began. *Turns out Trump just accused me of treason.*

The line was momentarily silent.

He did what?

Treason, I explained, and laid out the background to him.

I know Bill heard the anger in my voice. I ordinarily made a point of pride of being implacable, but as I stood there, I didn't care about the obvious fury in my tone. *The Bureau can't let this stand,* I argued. *Of course I have a personal stake in this, but the Bureau can't let Trump go making unfounded allegations of treason against agents who put their lives on the line for their country every day.* Since before Trump's election we had opted not to correct all kinds of slander against the FBI, expecting truth to win out; but it had only gotten worse. Trump had pushed the boundaries and, finding no sanction, pushed further. Realizing how angry I was, and recognizing that Bill was both my friend and my boss, I paused. *Look, I'm sorry,* I said. *I know it's provided fodder for them to weaponize. But Trump's behavior is out of bounds. Can the Bu say something?*

Let me track down someone in OGC and Public Affairs, he said. I'm sure Bill felt a range of emotions, but he sounded patient. *I'll call you back.* And then the line went dead.

My heart pounding, I walked back into the room where Aitan was waiting, and while we waited to hear back, we began drafting a statement. Even that was a difficult decision. Since Bureau policy prohibited me from commenting, the statement would need to come from my attorney. But how directly to take on Trump? Name him? Call him a liar? Show deference to the office?

In this moment Aitan and I faced a decision that I and countless other subjects of Trump's ad hominem attacks would face in the months ahead: how to respond to an immature, name-calling bully while still respecting the office of the president. We chose a solution that followed the old axiom "Never wrestle with a pig. You both get dirty, and the pig likes it."

Aitan and I opted for a succinct statement: "It is beyond reckless for the president of the United States to accuse Pete Strzok, a man who has devoted his entire adult life to defending this country, of treason. It should surprise no one that the president has both the facts and the law wrong."

As we tweaked words, my phone rang. It was Bill. I stepped back out of the room to speak privately in the elevator bay. Bill was with a deputy general counsel and the head of our Office of Public Affairs, he told me. He sounded subdued. I sensed that good news was not forthcoming.

I'm sorry, Pete, we're not going to say anything, he said. There wasn't much left to say. I felt awful. I sensed that they felt helpless, even embarrassed, and knew nothing was going to change. I told them that I was going to have my attorney issue a statement, then quietly thanked them and hung up.

I had a flashback to new-agent training at Quantico during a block of instruction about investigating violent crimes. The instructor, a senior agent who had spent the first decades of his career investigating gangs, told us that whenever possible we should always bring overwhelming strength and numbers to an arrest. There was no "fair" when it came to taking someone into custody. *A gang may be big or fearsome,* he said, *but you're a member of the biggest, baddest gang there is — the FBI. And we will always have each other's back.*

Until that gang is the president, I thought. Staring at the empty area outside the elevator banks, I had never felt more alone.

DUTY TO WARN

Trump's fury at me continued unabated. As of this writing, in less than two years he has launched more than 100 attacks on me in tweets, interviews, and other statements to the press. The president has variously labeled me "incompetent," "corrupt," "horrible," "hate-filled," "totally biased," "low," "terrible," "disgusting," "stupid," "a disgrace," "a total phony," "minion," "sick loser," "hating fraud," "fraud against our nation," "clown," "bad player," "phony," "dirty cop," "dishonest," "bad person," "sick, sick" person, "lowlife," "leaker," "liar," "psycho," "scammer," "con artist," and an "evil person." Trump has accused me of committing treason, claimed I had plotted a coup, and pronounced that I had "wanted to do a subversion." And when he needs a distraction, as on the day the House Judiciary Committee voted on articles of his impeachment, he rolls me out, going to his well-worn conspiracy of the insurance policy.

I did, and have done, my best not to allow the vitriol to affect me. But even at that early stage, back in January 2018, it took a toll. It's difficult to lose your anonymity, particularly the unexpected ways it affects you in public. Every time I stepped onto the sidewalk or into a coffee shop, I steeled myself for the possibility that some stranger might approach me and begin talking. Someone has to have strong feelings to approach you cold if they don't know you. And people had strong feelings about me — only strong feelings, it felt like. Gratifyingly, and humblingly, most of these feelings were positive in the people who chose to approach me. A much smaller number, less than 10 percent, were not. I felt like an odd personal bellwether for the administration's approval rating, as several times a week people came up to me to express support, encourage me, shake my hand or tell a story, or anonymously pay my bill at lunch — or, in the rare alternative, yell at me.

When I was with my family, I went out of the way to avoid placing ourselves in very public places that might bring polarizing attention. The hardest part was not knowing as people walked up what their feelings were, along with the worry that they might be in the tiny percentage of people prone to violence. Being in public felt like being in a bad part of town all the time. I felt like I needed to be constantly alert, watching people's hands for furtive movements, looking for odd shapes concealed along their beltlines and abdomens. Worse, my family had to do the same, even when I wasn't with them. It was exhausting.

The animus also began to directly affect my work in a way I never would have expected. Just before spring broke, my boss took the Human Resources leaders to Quantico to plan strategy for the upcoming year. We went for lunch and ate in the academy cafeteria, watching out the windows as a dreary rain fell and thinking glumly about the soggy commute home. As we walked back to our conference room, my boss pulled me aside. *Hey, Pete, one of our security officers is driving down. He needs to read you out of some programs.* That was security-speak for taking away my access to particularly sensitive categories of classified information.

It was an odd request. The administrative efficiency of our security folks was remarkably slow. The fact that someone was barreling down I-395 from D.C. to Quantico in an afternoon rainstorm for a bureaucratic process that took 30 seconds was strange, to say the least.

Can't we save the poor guy a trip and wait until tomorrow morning? I asked.

No, came the sheepish reply. *Wray's on the Hill tomorrow, and if asked, they want him to be able to say you've been read out.* In other words, fear was taking its toll on the Bureau again. These FBI employees feared that because of Trump's latest tactic of intimidation against those of us in the intelligence community, Wray would be crucified by committee Republicans if he admitted that I still had security clearance. He needed to be able to truthfully say that Peter Strzok no longer had access to certain classified information.

As irregular and absurd as this was, it eventually became clear that I was not the only one who faced this kind of petty and wasteful vindictiveness. Months later, over the summer of 2018, Sarah Huckabee Sanders announced that the White House had pulled the clearance of former CIA director John Brennan (though it's not clear to me that they actually followed the bluster with action) and was considering action on Comey, McCabe, me, and others. Trump was pulling every lever at his disposal to intimidate and discredit those he considered his enemies, regardless of the cost to national security.

The White House's public threats weren't the only thing on the news at this time. Conservative media played a willing accomplice as the texts burst onto the scene in January 2018. According to data assembled by Google, at one point I—just me, not including Comey or McCabe or Mueller or the general "no collusion, no obstruction" mantra—accounted for almost 15 percent of the combined daily airtime discussion on Fox and Fox Business News. It was deeply unnerving and uncomfortable being in the spotlight so much. Walking into a meeting on the seventh floor one day, Bowdich joked with me (when we were still joking), *I see you on TV more than I see my wife.*

Well, then, you really need to change the channel, I offered.

We chuckled, but not everyone was able to see the humor in all this absurdity, fear, and waste. That became increasingly hard for me as well. On

February 8, an itinerant bodybuilder named Cesar Sayoc began posting screen grabs of Fox News coverage of my texts and other conspiratorial content on his Twitter feed. He lived in a van plastered with pro-Trump and anti-Clinton stickers, while delivering pizza and occasionally working as a male stripper. A self-described Donald Trump "superfan," Sayoc populated his social media with photographs of himself at Trump rallies. As the spring of 2018 progressed to summer, his fixations and paranoid delusions deepened, fueled by his obsessive admiration for the president and the toxic political environment around him. His occasional mentions of me gradually shifted from mainstream cable news spots to retweets of unhinged conspiracy theories claiming links between me and Christine Blasey Ford, George Soros, and Hillary Clinton.

If not for what happened next, I might never have distinguished Sayoc from any one of the ever-increasing number of unbalanced and often vile extremists who populate Twitter, 4chan, and the dark reaches of the Internet. But every so often one of those people decides to take matters into his or her own hands. That's exactly what Sayoc did.

During the summer of 2018, pipe bombs began appearing in mail sent to former president Obama, Clinton, Clapper, and others associated with the far right deep-state narrative. Eventually fourteen were discovered. After an intense four-day manhunt, the FBI arrested Sayoc in Florida on Sunday, August 26, as he came off a job as a DJ at a male strip club.

Early the next morning, two agents — one from the FBI, the other from the Secret Service — arrived at my house. Finding no one home, one of them left a message on my cell phone. I called back and set up a meeting at the National Building Museum next to WFO. The museum, a soaring brick edifice inspired by Roman palaces, once housed the U.S. Pension Bureau. Inside, arcade galleries overlook a light-filled great hall, where I had spent many mornings sipping cups of largely undrinkable coffee at tables set up along the perimeter of the large open interior courtyard.

That was where the agents found me. They introduced themselves, and we walked up worn granite stairs to look for a quiet place to sit on the mezzanine. We found three chairs at a table overlooking the fountain in the middle of the great hall and sat. The agents didn't waste time.

You know what a duty-to-warn interview is, don't you? the FBI agent asked.

I do, I answered quietly. The "duty to warn" reflects the FBI's obligation to inform people when the Bureau develops information that they might be in physical danger. I had been on the opposite side of several such decisions but never on the receiving end.

You were on a list of people Sayoc was researching, they explained. The FBI believed that all the bombs he had sent were accounted for, but they urged me to be on guard for suspicious packages nonetheless. I didn't mention that it was the second time I'd heard that advice recently.

I asked the agents if they had any more information, but they didn't. The whole unsettling exchange took less than 10 minutes. My sense was that the Bureau was acting as quickly as it could to get word out to those affected, and I was thankful.

One of Sayoc's attorneys would later explain in court that his client's blind admiration for the president had fed his spiral toward violence. "It is impossible, I believe," the lawyer said, "to separate the political climate and his mental illness." I had to agree. After investigating Trump's assaults on the weak links of government institutions, I was again experiencing his impact on the vulnerable fringes of American society, on troubled people susceptible to suggestion and inflammatory rhetoric.

In those moments when I lay awake at night, during this period and in the months that followed, I often would wonder how we got here. Growing up in troubled countries abroad, I had always believed that I could recognize the symptoms of internal deterioration, of how the middle collapses without a strong spine of democratic institutions. And now I was seeing something else that I had experienced overseas when governance fails. Authoritarian leaders and tin-pot dictators don't tolerate dissent or criticism, and when they hear it, they smear their critics in outlandish terms, as traitors, as enemies of the people, as saboteurs and spies. If they can imprison or execute their critics, they frequently do. If they can't, they call for their imprisonment or execution instead, or demand mob retribution against them. Corrupt and compromised, with no moral center and no ethics and only their own self-interests to guide them, such leaders see criticism as a challenge to their legitimacy and, when challenged continuously, rage louder — ruining lives, destroying careers, and worse.

Now, in the United States in 2018, that was happening to me.

16

"That is who we are as the FBI"

ON MARCH 16, just over a month after I'd visited DOJ to review my texts, the department took another step that served as an unmistakable warning for those of us in the FBI who, in the execution of our duty, had angered the president. That day the attorney general hammered another one of Trump's perceived enemies — and another link in the FBI chain of command — out of the Bureau. In much the same way that he and Rosenstein had given Trump the pretext for terminating the FBI director, Sessions now fired McCabe, the deputy director of the FBI.

McCabe was the subject of an ongoing disciplinary process, although at the time I didn't know the specifics of it. But I did know that Sessions had cast McCabe out of the Bureau in an unprecedentedly accelerated disciplinary process, and just hours before he was eligible to retire and begin drawing his pension.

To me, the unheard-of pace of McCabe's termination smacked of improper political intervention. Not until later would I learn the full backstory to his termination. In November, the IG had informed Wray and Bowdich that he would be issuing a separate report about McCabe and his statements to the FBI and the IG about whether he had authorized a release of information to the *Wall Street Journal*. The IG would eventually conclude that McCabe lacked candor in several of his statements to investigators. Wray had told McCabe that he could not remain as deputy director. Rather than be reassigned, McCabe had taken leave until he was eligible to retire. But that gambit, used before by countless soon-to-retire agents who found themselves under disciplinary scrutiny, had failed.

What's more, while McCabe's termination didn't seem as problematic from a counterintelligence perspective as Comey's had, the context in which it occurred had gotten much more troubling. Between the firings of the top FBI officials, as Trump's attorney general had continued hacking away at the FBI, Trump kept up his unrelenting attacks against anyone connected to the Russian investigation while also continuing to ingratiate himself to Putin.

Two days after McCabe's firing, Russians elected Putin to a second six-year term. Shortly after, Trump called Putin. Trump's staff gave him a note with simple, all-capital letters: "DO NOT CONGRATULATE." Trump heartily congratulated Putin anyway, while failing to take him to task for the Russian government's attempted assassination of Sergei Skripal, just days after the British government identified Russia as the culprit behind the bungled effort. Trump's jocular behavior stung doubly for those of us in the intelligence community. Skripal was one of the former Russian intelligence officers whom our government had extracted from Russia in exchange for the return of Andrey, Elena, and the other illegals we had painstakingly tracked down as part of Operation Ghost Stories.

To cap off the parade of horribles, a week later the DOJ inspector general, Michael Horowitz, bowed to calls from conservatives in Congress by announcing that he would begin an investigation of several early elements of Crossfire. The review would include the Carter Page FISA warrant; Christopher Steele, who had produced the now-infamous "dossier"; and Bruce Ohr, a career DOJ official and Russian organized crime expert who had known Steele for years and had remained in contact with him as the Bureau operated and then closed Steele as a source.

The political sands were shifting, and people were running scared. Under attack, on alien terrain, public servants and organizations that were supposed to be immune to political pressure were caving in to that pressure or at best withdrawing into defensive postures, particularly within law enforcement agencies. Of course, every U.S. president has the responsibility and authority to set policy agendas, whether in foreign affairs or in domestic law enforcement. But Trump was blatantly overstepping this prerogative. On an unprecedented scale, he was attacking legal investigations of possible violations of established law. Even more disturbing, political appointees and senior public servants whose jobs were nominally to uphold the law were succumbing to his demands to avoid a withering Trump tantrum.

Sadly, by this point I had come to expect the attacks from Trump and his partisans on the investigation; what surprised me was the lack of a spirited defense, particularly from within DOJ and the FBI. Some of the organizational silence was undoubtedly a move away from a perception that Comey had been too publicly engaged. Wray had a very different leadership style from Comey, and made clear his "plow horse, not show horse" preference. But that didn't explain the whole picture.

FBI senior executives seemed now to be assiduously avoiding any knowledge of, let alone involvement in, the Crossfire cases. At a seventh-floor briefing about a Human Resources issue in early spring 2018, I heard a senior leader joke about a question he had gotten during an FBI town hall about "manifold." I had no idea what he was talking about until he explained that he was surprised FBI employees were so interested in what Mueller was doing, at which point someone corrected him that it was "Manafort."

The event bothered me, because I really liked and respected (and still do) the executive telling the story. He was bright, thoughtful, loved the Bureau and its mission, and was a great leader. But he knew nothing about Mueller's work and seemed content to keep it that way. And really, who could blame him?

By this point, Trump and his enablers at DOJ had made it clear that any FBI official who got involved in the Russia probe would pay a high price. Comey had been fired, McCabe had been fired, and I was being savaged in the press. The Bureau's leaders had strong personal and organizational reasons to remain silent and unaware, to avoid the withering ire of Trump and his supporters. Illustrating this atmosphere, Wray admitted in congressional testimony following the conclusion of Mueller's work that he had not read all of the Mueller Report. Trump had succeeded in turning Crossfire into a third rail.

Midyear too had been a liability for any FBI investigator or official who had participated in it. In any other time, Midyear would have been seen as the consummate independent FBI investigation done by true professionals. But in the age of Trump it had become toxically politicized. Starting with the "lock her up" chants and Trump's later repeated false allegations that Clinton had somehow been given a sweetheart deal to avoid prosecution, association with our efforts could result in being attacked and sidelined.

The case investigators who had worked on Midyear and Crossfire felt some of the most perverse leadership impacts. At one point an agent who had been on Midyear reflected that Wray felt like a "step-director" and our team the proverbial redheaded stepchildren. DOJ and FBI leaders appeared to have embraced the idea that Midyear and Crossfire had been the isolated work of a small group at headquarters — an attitude that ignored the facts that Midyear had involved hundreds of personnel from across the FBI and Crossfire had been pursued by investigators in three of the FBI's five largest field offices, under the watchful eye of each chain of command. These cases, which initially had been fully supported at the highest levels, were now being disowned by the officials at those same levels. The professionals who had been involved in these investigations were being shunned and disavowed.

However indefensible, the short-term message was clear: DOJ and the FBI's new leaders were disclaiming responsibility for any investigation relating to Midyear or Crossfire. The long-term message was far worse: it was just too perilous to investigate matters relating to Trump. The Bureau's top officials seemed to be telling their subordinates, in essence, to stand down, not to do what we had always done — our work, irrespective of who it involved.

Fortunately, the sense of duty of the vast majority of people at the FBI, and in the law enforcement and intelligence communities generally, was still guided by their deep commitment to the Constitution and the rule of law. For such men and women, these ominous signs served not as a deterrent but rather as a challenge to work harder than ever before.

BAIT AND SWITCH

As Trump continued to attack me and the other unlucky FBI officials who had become the faces of the Russia investigation, the DOJ Inspector General's Office was still reviewing the Bureau's handling of Midyear. Throughout the winter and spring of 2018, whenever asked, I made repeat trips to the IG's offices on New York Avenue near the White House to walk the investigators through the minutiae of what we had done on Midyear. Sitting in windowless conference rooms for hours, the investigators were exhaustive as they poked and prodded each of the decisions that we had made in the case: for instance, our use of subpoenas and search warrants, our planning of interviews, and our interactions with DOJ. They also continued to scrutinize my texts, which I also explained to them in detail.

It was important to me that the IG have the facts about the Clinton investigation — all of them. I believed that if they heard the full story and were the honest brokers that IGs are intended to be, their conclusions about Midyear Exam would mirror my opinion of it: that talented and hardworking agents, analysts, and professional staff had conducted an extraordinarily thorough investigation; that our actions throughout clearly demonstrated that no biased act ever did or could occur; and that anything in my texts didn't change one bit of that. The IG investigators took advantage of my willingness to help despite the personal costs, repeatedly relying on my memory and notes to form, and then review and edit, the factual backbone of their report's narrative.

In time I actually found myself looking forward to the public release of the IG's report. By late May the IG had concluded in his draft report that there was no documentary or testimonial evidence of bias in Midyear Exam. After months of deep and thorough investigation, he also saw no significant issues with any of our major investigative decisions. What's more, his investigators noted that I had been one of the most vocal advocates for aggressive investigative techniques against Clinton — a finding that seemed sure to deflate the countless conspiracy theorists who thought I had plotted against Clinton's opponent in the 2016 election.

The reality is that my aggressiveness wasn't aimed at or limited to Clinton. Good agents are aggressive, and successful cases are the products of tenacious and probing investigative work. Still, I was heartened that the IG's office came to the conclusions it did, and relieved that it finally saw that the investigation had been an evenhanded process.

Those warm feelings toward the IG didn't last long. A little more than two weeks later, I was back in the IG's conference room with Aitan. This time I was furious. In the interim, the report had been sent to DOJ, which meant that its conclusions likely had been seen by the White House too. And now some of those conclusions had been changed.

The two primary IG investigators sat across from us, and Horowitz's disembodied voice joined us from a triangular speakerphone on the center of the table. The IG claimed that he had made changes to the report after a contractor hired to do a deep forensic dive to look for additional texts had recovered additional material. Although the new texts were substantively similar to the ones they had already reviewed, Horowitz no longer seemed willing to take what could be a costly political stand by finding that we had acted objectively and professionally.

There was one exchange in particular that seemed to unnerve Horowitz, making him second-guess the thoroughness of his team's investigation and conclusions. Days after Trump's repeated attacks on the immigrant family of Humayun Khan, an army captain who gave his life for our country when he was killed by a suicide bomber in Iraq in 2004, I had expressed my view that Trump would never be elected. In a text that to this day I still do not remember writing, late at night I wrote, "We will stop it." It was an artless comment but conveyed my firm belief that Trump — a TV personality who was proving to be a crass, racist, misogynistic force in our national politics, and who was espousing views that radically undermined our national security commitments to NATO and the world — would not be chosen by the American people to become president.

The discovery of this text triggered Horowitz's apparent risk aversion and caused him to make a fateful and ill-considered edit to the report he had already drafted. He should have investigated this new evidence to determine whether or not my comment suggested anything different from what he had already concluded—especially given that my comment had nothing to do with Midyear and that he was going to continue his investigation into Crossfire. But Horowitz decided instead to change a core conclusion about me in the report. Given the Trump camp's pressure to get the report out and to reach conclusions that were critical of the FBI along with the time pressure to release his findings, maybe he didn't feel he could take the steps necessary to satisfy himself that the new texts didn't change his original conclusions.

Ramping up the critical tone of the report, he addressed his fear by adding a new assessment that he couldn't prove a negative—that is, he could no longer eliminate the possibility that bias had played a role in the delay of several weeks between NYO's discovery of Clinton email on the Weiner laptop and our eventual procurement of a search warrant for it. Horowitz also believed that we had prioritized Crossfire over Midyear, which—in light of this new text—he took as evidence of anti-Trump bias on my part.

The IG's rush to judgment was infuriating to me for a number of reasons. Perhaps most frustratingly, the change ignored the facts and seemed to be based on impulse. As I described in Chapter 8, the delay of the Weiner warrant was the result of an honest miscommunication among the agents and attorneys who were assigned to follow up on that lead. More than a half-dozen people had been involved in that process, including those primarily responsible for following up, and the IG found no fault with any of their motivations. Most significantly, Crossfire was inarguably a much more urgent and consequential investigation, so even if we *had* prioritized it over Midyear, we would have been justified in that decision. If the IG would only take the time to look into this and the related issues, I thought, he would see that they had no bearing on his initial findings.

Adding to my aggravation was an earlier request from one of the two primary investigators: *If you don't mind, we'd really appreciate it if you could read through the whole report for accuracy. Ordinarily the Bureau gives a methodical scrub of our reports, but we don't think they're going to do that this time.* Most draft IG reports are written with the expectation that they are the beginning of a process to allow affected agencies and individuals to correct inaccuracies and missed facts, rebut points, and adjust language to present a balanced view. Not this time. With Comey gone, McCabe gone, and Trump raging, the Bureau's leaders wanted nothing more than to put the episode behind them. They were willing to take what the IG wrote, accept it wholesale without comment, and try to move on as quickly and as quietly as they could. There is a reason for those routine reviews, to correct errors, explain misconceptions, and address faulty assumptions; that benefit was lost here. The IG's investigators knew it, and admitted as much to me in the hallway outside their office. In the FBI's absence, the IG's staff was asking me to fill that void.

Barely controlling my anger, I factually rebutted their analysis. I explained that within two hours of learning about the Weiner laptop, I had assigned a supervisor at WFO with no role in Crossfire to follow up with New York. He did that with his agents and analysts the next day, reporting back to me in less than 24 hours that New York was still in the process of forensically processing the laptop. Those were hardly the actions of someone trying to back-burner something, I pointed out. Just like all the WFO personnel (whom the IG *didn't* fault), I appropriately believed NYO would notify WFO in the normal course when the processing was complete, so I returned to the hundreds of competing and sometimes more exigent issues demanding my attention. The simple fact is that everyone who was working on the matter believed the ball was in someone else's court — an unfortunately common error in overtaxed bureaucracies.

I also rebutted the IG's assertion that I had put more emphasis on Crossfire than Midyear. There was simply no equivalence between Midyear and Crossfire. At its core, Midyear was a glorified (because it involved Hillary Clinton) mishandling case, the type of case that but for the celebrity subject rolls through WFO on a weekly basis. Crossfire was a first-of-its-kind, enormous investigation into complex, ongoing attacks on our presidential elections and Russian interactions with members of one of the candidate's campaigns. *Of course* I had prioritized that.

As I spoke, the two IG investigators scribbled furiously on notepads in front of them.

Pete, Horowitz said, *we're not saying you acted with bias. We're saying we can't eliminate the possibility that bias played a role in your decision-making.*

How in the hell do I prove a mental negative? I thought. I pointed out that they had access to everything and anything they wanted but that in their months-long comprehensive inquiry — involving thousands of hours of investigation, millions of pages of documents, and hundreds of interviews — the IG's team had found no evidence of bias. In fact, that had been their conclusion only days before. Now, based on fear that they might have missed something, they seemed to be using a fabricated concern — the text they claimed to be worried about had been sent months before we even knew about the Weiner laptop, and in any event was irrelevant to Clinton — to protect themselves from Trump's wrath.

So I went through, again, what I had done with the laptop. If there had been any biased act, I argued, they would have found evidence of it, but there hadn't been any, because it hadn't happened. They agreed.

In fact, I argued, all the facts they had developed pointed to the *opposite* conclusion: if anything, my actions *hurt* Clinton and *helped* Trump.

I don't know what more I can say, I said, shaking my head. Looking at the two investigators, I finished: *You reviewed the file and each and every text, all the email, each decision, and every action. From all of us. You know how many interviews you did. You know there was no evidence of bias, because it doesn't exist. If it did, you would have found it. The fact that you didn't is the point, not some intangible concern that you might have missed something.*

None of it made any difference. The IG had made up his mind. I left feeling indignant, defeated, and persecuted.

The IG released the revised report on Thursday, June 14. I felt then, and still feel, that it was the most irresponsible kind of character assassination

and perpetrated for the lowest of reasons. Horowitz appeared to be unnerved by how the texts related to what was coming next: his upcoming
investigation into the Crossfire probe. So unnerved, it seemed to me, that
the discovery of these new texts during his Midyear investigation made him
unwilling to consider evidence conflicting with his conclusion, as if that
evidence didn't exist or didn't matter. Whether his conclusion was motivated by a fear of having missed something or fear of Trump and his supporters' rage if he failed to back up their narrative, I can't say. But clearly
he was going to express his outrage regardless of the fairness, impact, or
cost.

Predictably, Trump gloated. "I am amazed that Peter Strzok is still at the
FBI, and so is everybody else that read that report," he said the next day
from the north lawn of the White House. "And I'm not even talking about
the report; I'm talking about long before the report. Peter Strzok should
have been fired a long time ago, and others should have been fired."

It sure felt like the Bureau was listening. That Friday afternoon, the day
of Trump's tweet, I received word from the Bureau's Office of Professional
Responsibility (OPR) that I had been proposed for termination and was
being placed on administrative leave while the adjudication process moved
forward. Just as McCabe had been fired in record time, mere hours before
he was eligible to retire, I was on the chopping block within hours of the
IG's releasing his report. I packed up my things that Friday evening and
headed home indefinitely to process the unthinkable possibility of losing a
job I loved and the career I had worked so hard to build.

It is hard to explain how unusual this accelerated process was. The FBI's
disciplinary system is notorious for its unreasonably long delays, which tie
up agents in administrative limbo while they wait for their cases to be resolved. Not this time, though. Not with Trump raging against me and his
howls bouncing around the echo chamber of Fox News.

The next week Sessions went on a radio show in Boston and announced
that I no longer had a security clearance. (He was wrong — although I had
been read out of certain programs, my clearance was otherwise unaffected

—but I have to assume his comment was intended for an audience of one, so the truth didn't matter.) Back in D.C., meanwhile, Horowitz proceeded to the next and final phase of his Midyear probe: after releasing the updated report, he needed to testify about his findings to Congress.

As I'd expected, the IG's congressional testimony started out along partisan lines, but midway through, Horowitz strayed from his report, going beyond its findings to further vilify me and others. I was glad to hear a defense of our work but disappointed to watch Horowitz—wittingly or not—be led by Trump-enabling congressmen into an alternate universe of conspiracy theories, in which the Midyear, Crossfire, and now Mueller investigations were all part of a biased deep-state coup. Horowitz's team hadn't yet investigated Crossfire, but that didn't stop him from speculating about it, making qualified statements like, "I think a reasonable explanation of it or reasonable inference . . . is that [Strzok] believed he would use or potentially use his official authority to take action" against Trump. The inspector general didn't bother to mention that his investigators had tried —and tried, and tried—but ultimately failed to uncover any actions on my part that supported this conclusion. Nor did he mention, more accurately and relevantly, that his office had accumulated a significant amount of evidence to the contrary.

Horowitz's gratuitous and unsupported comments poured gasoline on the fires that were burning out of control on the fringes of public opinion, helping them to creep into the mainstream. By going beyond and distorting the findings and scope of his report and letting his fears and moral outrage taint his conclusions, the IG fueled a vast conflagration of conspiracy theorizing. In so doing, he allowed his office to be misused by people who were hell-bent on attacking Midyear and Crossfire because undermining our work advanced their agenda—one they were willing to pursue regardless of reality, and regardless of the harm they might cause to our country or any individual.

I burned with indignation, and, as I had so often in recent months, I yearned to speak out. Soon I would get that opportunity.

MY TURN

On June 15, the Friday before Horowitz's testimony, the Republican-led House Judiciary Committee announced that it was beginning the process of issuing a subpoena for my testimony. When I heard that, I rolled my eyes. The move was a public relations stunt, pure and simple, intended to make me look like I had something to hide. My attorneys and the Bureau's Office of Congressional Affairs had been working with Congress for months to set up a date for me to testify. I was more than willing to testify; I wanted to.

I knew that the committee was merely trying to whip up drama, but it struck me that it wasn't being smart about it. Sure, subpoenaing me made me look guilty of something and added intrigue and interest to the narrative that congressional Republicans were trying to build about alleged bias in the 2016 investigations. But the members of the House Judiciary Committee majority seemed oblivious to the possibility that the truth about my desire to testify would come out. More importantly, they didn't seem to be concerned that the true motivations for these inquiries might become clear. I believed that the truth would emerge during testimony, hopefully causing six months' worth of conspiracy theories to collapse in the eyes of objective observers.

My first turn in the barrel was on Wednesday, June 27. One of the administrative absurdities of the position in which the government had placed me by suspending my clearance became evident little more than 48 hours before, when the Bureau straightened out the administrative approvals to read me back into my clearance so that I could review my notes in advance of my testimony. After being cut off from my casework for more than ten months, I had less than two days to prepare for what was sure to be a thorough — and thoroughly hostile — grilling about it.

On Wednesday morning I met my attorney, Aitan, and we hopped in a cab to the Au Bon Pain across from FBI headquarters to grab a cup of coffee and wait for our ride to Capitol Hill. Soon a white van containing a phalanx of Bureau folks — two attorneys from the Office of General Counsel and two members of our Office of Congressional Affairs — appeared on the ramp leading up onto 10th Street from the bowels of the Hoover Building. We climbed in and the van pulled out onto Pennsylvania Avenue, heading toward the Capitol.

My first congressional testimony was to take place in a closed setting in the Rayburn House Office Building. It would turn out to be a sedate affair compared to the open session to come. Unlike the public, televised hearings that followed, in which each member of Congress got a five-minute allotment of time to try to score points by asking or pontificating about whatever he or she wanted, the closed session would be a series of hour-long sessions alternating between Democrats and Republicans on the committee, conducted by a mix of members and staffers. Unsurprisingly, Trump was not happy with the closed hearing, tweeting, "The hearing of Peter Strzok and the other hating frauds at the FBI & DOJ should be shown to the public on live television, not a closed-door hearing that nobody will see. We should expose these people for what they are — there should be total transparency!"

For once, I couldn't agree more, I thought. *The president is right: let the truth come out.*

My testimony was secret, but my entrance was not. As the group of us walked through the outer doors, I heard a shout: "There they are!" Looking past the security checkpoint, I saw a swell of cameras and reporters surging toward us. It was the first time I had seen this. I remembered how Comey described to the Senate Intelligence Committee the massed media as being like the chaotic crush of feeding seagulls. Once through security, we were in their midst.

Ignoring the clamor around us, I turned to one of our congressional liaisons. *Where are we headed?*

He nodded toward a hall. *This way,* he said, and we began moving in a thicket of lights and microphones toward the hearing room. I recognized one of the reporters in the crush as the reporter with the rage-filled face after Comey's speech. As she kept pace on the periphery of our procession, she was the only one I noticed continually trying to thrust a microphone toward me.

The interview was exhausting, but it was a cakewalk compared to my subsequent experience on the Hill. While Republicans and Democrats took hour-long turns for more than ten hours, the questioning also varied between members and their staffs, which made it disjointed. Because there were no cameras, it felt like there was less preening for public consumption, and because there were effectively no time limits, there was the opportunity for a dialogue between me and my questioners, rather than five minutes of badgering exposition or grandstanding.

That's not to say that things weren't contentious. The members of Congress whom I expected to be partisans out to twist my words to suit their predetermined ends rather than to find out the truth didn't disappoint. At one point, during an excruciatingly long and characteristically pedantic exercise, congressman Trey Gowdy attempted to parse one of my text messages word by word. Aitan interjected, "Congressman, you have repeatedly and publicly talked about how you want to hear from Agent Strzok. It now appears that you don't want to hear his answers, you want to hear your questions and then cut off his answers so that he can't give them." I got some fleeting satisfaction from Aitan's comment, but it wouldn't be the last word. The presence of TV cameras during my later public testimony would bring out an even more theatrical side of this particular representative.

As the hours rolled by, I realized that I could hold my own with the questioners. For the most part, the staff asking the questions were well prepared and at least competent. The members of Congress themselves — well, they varied. As it turned out, the interlocutors who were most in Trump's corner were less effective, in large part because I had the benefit of truth on my side. By afternoon I felt invigorated, which was good, because if there is one thing that testifying on Capitol Hill has taught me, it is that you cannot let your guard down.

I thought we would finish in early evening, but by the time evening arrived we still weren't done. The only questions I hadn't answered were the ones the FBI had told me I couldn't. Still, several Freedom Caucus members wanted to pursue far-fetched theories about the FISA process and other classified non sequiturs, so we adjourned to a secure room for classified discussions. We were walking down the hall to our new room with a Capitol Police buffer when the media throng descended on us again, trying to get a comment. I ignored them, feeling the fatigue of having been constantly alert to every small detail for 12 hours.

As we sat down in the secure conference room, I wondered how long this would continue. Looking across the room, I noted a similar fatigue on my inquisitors' faces. That was the moment that I realized I would accomplish what I had set out to do that morning. Just as the army had taught me to cope when the finish line unexpectedly moved off into the distance, I knew I could persist. *Let's go,* I thought, and — allowing myself a bit of gamesmanship and humor — I suggested to the group that we get some bourbon to make the evening easier.

We didn't finish until close to 10 that night. By the time the classified discussions wrapped up, I had reached the point at which I really did want some bourbon. But the Bureau held one last card of absurdity in its hand before I would be allowed to go home. *We need to go back to headquarters to read you out,* one of the group said.

Are you fucking kidding me right now? I thought. Instead I replied, *Look, can't we do this tomorrow?*

Sorry, no. A couple of security folks have been standing by at headquarters. We can read you back in tomorrow. It made no sense to argue — I wasn't talking with the people who had made the decision.

After a career of being entrusted with the most critical national intelligence secrets, I can't imagine that the FBI was afraid of what I might do overnight with an active security clearance. More likely they were trying to hedge against criticism that I hadn't been immediately been read out and therefore hadn't been treated in the same manner as others, perhaps Comey and McCabe, who likely faced similar circumstances. *Those poor security guys,* I thought. *They've been sitting there for hours with nothing to do for no good reason.* It was actually worse than that — they were swept up in a silly dance of waiting for grandstanding politicians to waste the nation's national security resources for partisan pursuits while being kept from doing the FBI's real job of protecting America.

I could do little more than shake my head at the Catch-22 level of bureaucratic absurdity. Our little group walked across Independence Avenue to climb back into the white van, and trundled back to headquarters for the three minutes of unnecessary paperwork. Then, finally, I got to go home. I enjoyed the bourbon.

"IT'S GOING TO BE UNPLEASANT"

I was deeply asleep when Trump's reviews of my performance came in. They were, of course, predictably harsh, threatening, and inaccurate.

At 4:02 in the morning, the president took to Twitter to say that I had been "given poor marks on yesterday's closed-door testimony and, according to most reports, refused to answer many questions. There was no Collusion and the Witch Hunt, headed by 13 Angry Democrats and others who are totally conflicted, is Rigged!" Pausing for an hour and a half, he returned to the topic at 5:30 a.m. to announce, "Peter Strzok worked as the leader of the Rigged Witch Hunt for a long period of time — he got it started and was only fired because the gig was up. But remember, he took his orders from Comey and McCabe and they took their orders from you know who. Mueller/Comey best friends!"

How can the leader of the free world have time for this? was my first thought when I saw the tweetstorm — not the last time I would ask myself that question. It was surreal to see Trump following my testimony so obsessively, like an itch he couldn't scratch. It occurred to me that he was trying to stop something that he didn't — or at least shouldn't — have any power over.

Trump's reviews notwithstanding, I thought the session had gone well. Congressional Republicans confirmed that impression when they immediately moved to block calls for the release of the transcript to the public, which they succeeded in doing for months. For a moment I thought my time on the Hill had ended and that I had done well for the FBI, and for my teams in particular. But that hope was dashed when Judiciary staffers reached out about a week later to begin scheduling another appearance for me, this time in the open hearing that I had originally sought and that Trump claimed to want.

I reminded myself that this was what I wanted. Based on what I had seen during my prior testimony, I knew it was going to be an even bigger circus. What's more, because it was a joint hearing between the Judiciary, Oversight, and Government Reform Committees, there were going to be far more members than seats in the usual hearing room. So the committees found a larger space in the same building, one used by the House Armed Services Committee.

Kate Duval, one of my attorneys, knew Congress inside out and managed to get us into the hearing room to take a look in advance. The room held several semicircular rows with tiered seats for members. On the opposite side, seals of the armed services lined the back wall above oil paintings of former committee chairmen. As I walked around the chamber, she explained that although it held more members than the committees' normal space, it still wouldn't be large enough. So many members planned to appear that after their five minutes to ask questions, they would depart to give up their seats for those waiting, a round robin of inquisitors that was sure to continue throughout the day and likely into the evening again.

I made my way over to the witness table, where I would sit at the upcoming hearing, and moved away all the chairs away except one. I slid it to the middle of the table and sat down, trying to imagine what it would feel like once the room was filled to bursting with politicians, their staff, media, and curious onlookers. Between the witness table and the members' seats there was pit space that, Kate explained, would be filled with photographers when we entered the chamber. Once the hearing began, they would be moved off to the side, leaving the fixed video cameras to cover the proceedings. At the moment, though, the hearing room was cavernous, silent, and empty.

It's going to be unpleasant tomorrow, one of the attorneys offered sympathetically.

I know, I responded. *But at least I'll finally be able to speak my piece publicly and defend the FBI and our work,* I thought to myself.

The next morning I carefully picked one of my favorite striped ties, with a red, white, and blue pattern, and put two copies of my opening statement into a leather briefcase, along with a pen and a pad of paper. The first copy of the statement was printed in a 20-point font to make it easy to read, and its corners were worn from the many times I had reviewed it, again and again until it said what I wanted it to.

That wasn't all I put into the briefcase. In an inside pocket I tucked a letter from a retired air force officer I had never met, encouraging me to speak the truth simply and clearly, including the crux of Crossfire: that we were investigating whether Americans had conspired with a hostile foreign nation to intervene in our presidential election. I knew about the issues at the heart of the Russia investigation, of course, but the letter was a heartening reminder that people across America did too. I wanted a reminder of support, a lodestar, close at hand during the testimony. It would help me to navigate the rhetorical minefield laid down by the president, who in the days leading up to the testimony had repeatedly announced himself a victim of a "Rigged Witch Hunt," and who I was certain would follow my testimony — and react loudly, and negatively, to it.

I had another lodestar, should I need it. My wife had told me that if I got angry at any of the questions pelted at me during the hearing, I should pause for a beat and spin my wedding ring before speaking.

The hallways of the Rayburn House Office Building were even more crowded than they had been during my first appearance. The hearing room was now packed, with people lining the walls and a scrum of photographers crouched in the pit in front of the witness table. As my companions and I stepped into the hearing room (beyond the two people from my law firm, the Bureau had sent an entourage of four to monitor the proceedings), cameras pivoted toward us with the staccato clicking of shutters. From the youthful appearance of the people lining the halls and the walls of the hearing room, my guess was that members — House Republicans, I assumed

— had placed their staff and interns along the route we were walking to amplify the spectacle for the cameras.

Our group of seven made its way down to the witness table. Aitan took a seat behind me, with the four Bureau attorneys and congressional liaisons filling in the seats behind me to my left and right. As we got settled, photographers jostled to snap pictures of me — images I had worked so hard to avoid over the course of my career.

I opened my briefcase, removed the portfolio, and arranged my opening statement, pen, and pad. I shifted small water bottles — compliments of the committee — to a spot where they weren't in the way. Then I waited. Chairman Bob Goodlatte was late. I briefly wondered if it was a deliberate but ultimately misguided effort to ice the witness. I spent the time chatting with Aitan.

After the chairman arrived, the hearing almost immediately descended into partisan acrimony. Members talked over each other, shouted out points of order, and demanded to be recognized. I glanced down at my watch, captive to their posturing and mutual hostility. We were nearly 30 minutes in and I hadn't given my opening statement, let alone answered a single question.

When the questioning began, I felt an immense sense of relief. After six months of silence while under constant attack, I was finally able to address the substance of what we had been investigating. In my opening statement, I directly laid out to the committee — and the American public — what we had done and not done. "I testify today with significant regret," I began, "recognizing that my texts have created confusion, and caused pain for people I love." But I defended my right to have personal political opinions, and noted that if I had truly intended to hurt Trump's candidacy, I would have revealed the investigation and its substance. Given my privileged role in investigating the connections between his campaign and Russia in the months leading up to the 2016 elections, such a revelation would likely have been campaign-ending. But of course I hadn't and never would have done any such thing.

More importantly, I explained, the Russia investigation was not at its core about Trump: it was and always had been about the threat from Russia. "I understand we are living in a political era in which insults and insinuation often drown out honesty and integrity," I said. "But the honest truth is that Russian interference in our elections constitutes a grave attack on our democracy. Most disturbingly, it has been wildly successful, sowing discord in our nation and shaking faith in our institutions. I have the utmost respect for Congress's oversight role, but I strongly believe today's hearing is just another victory notch in Putin's belt, and another milestone in our enemy's campaign to tear America apart. As someone who loves this country and cherishes its ideals, it is profoundly painful to watch, and even worse to play a part in."

I hoped that the underlying message of my opening remarks was clear: the recent weaponization and purposeful misreading of my texts was not merely an affront to me and to the FBI. In their attribution of bias in the opening and pursuit of the investigations of Russian interference, these cynical political gambits constituted a dangerous diversion from a real and ongoing national security crisis, one to which current political leadership seemed unable or unwilling to respond.

Almost immediately Gowdy went on the attack about my alleged bias. When I was given the opportunity to respond, I took a deep breath before starting to answer. And then I told him the truth, just as the letter in my portfolio encouraged me to do.

Russia, I explained, had interfered in the most fundamental act of our democracy: our casting of votes to select our leaders, in this case the highest in the land. The investigative question, I continued, was not who voted for whom but if and how Russian influence had skewed that process. I never, ever — under oath, *ever* — allowed bias to influence my work. Like everyone else, FBI agents have personal beliefs, but we check them at the door every day when we walk into work. That is our code, what we hold each other to every minute of every day.

I was almost through, but there was still one more thing I wanted to say while the cameras were rolling. My colleagues at the Bureau, I said, my voice rising, "would not tolerate any improper behavior in me any more than I would tolerate it in them. That is who we are as the FBI. And the suggestion that I, in some dark chamber somewhere in the FBI, could somehow cast aside all of these procedures, all of these safeguards, and somehow be able to do this is astounding to me. It simply couldn't happen. And the proposition that that is going on, that it might occur anywhere in the FBI, deeply corrodes what the FBI is in American society, and the effectiveness of their mission, and it is deeply destructive."

It felt like I had opened up the pressure release valve after silently watching for months. I was drained but also unburdened, and it took a moment for me to register what was happening around me.

The room had erupted with clapping from the gallery and cheers and pounding on and underneath the tables from the minority side. The Republican majority sat silent.

From that point on, the majority's members increasingly filled their time with statements and assertions rather than questions, preventing me from correcting their misstatements and challenging their false accusations. When I attempted to respond, the chairman told me I could not, since a question had not been asked. *Great,* I thought, *Alex Trebek's* Jeopardy! *rules of parliamentary procedure. Lies can be lies, so long as they're not couched in a question.*

In a day filled with moments of absurdity, a particularly aggravating one came during a break immediately following an exchange with a congressman about my texts. As our group filed into a small room off the side of an antechamber behind the majority side, one of the attendees from the FBI's Office of Congressional Affairs remarked sympathetically about the abuse I had just experienced. He noted that he never wanted to be caught in such a predicament, and that that was the reason he used his personal phone to communicate with the head of his office about some work matters. Hearing this, one of the Office of General Counsel attorneys froze, eyes wide, jaw dropped. The FBI staffer had just proven the absurdity of the feigned partisan outrage about what was truly ubiquitous behavior.

What am I supposed to say to that? I wondered to myself. I thought for a brief second, then said, *Look, I think based on everything you're seeing here today, that's actually a bad idea. And you probably shouldn't be doing it.* I felt bad for him, because he seemed nice and genuinely didn't see the issue in what he was doing. It was also aggravating for me to hear, because as I and everyone else at the FBI knew, some of the behavior being held out as outrageous was hardly rare, if not commonplace.

We returned to the session, and as the afternoon went on, the questioning continued based on seniority. I learned some things — for instance, at least one congressman trained as a dentist believed that his profession conveyed an expertise in reading body language — and I had my patience sorely tested as some members stooped to ad hominem attacks and brought my family into their questioning. While the Bureau had taken my official smartphone, many friends had my personal number, to which they sent their support throughout the day. It is hard to overstate the positive impact those messages had on my morale. At one point late in the afternoon, I looked down as we were about to begin again to find an article forwarded from the satirical website *Onion* about the day's proceedings and laughed out loud.

Although the proceedings were the best opportunity I would have during this difficult period, to defend myself, my teams, and the Bureau, they were nevertheless a travesty. The committees seemed focused more on scoring points than on conducting legitimate oversight. Time, resources, and attention — on the committees' side as well as the FBI's — were squandered with spectacles such as this one: show trials of experienced investigators who, in the course of their duties, had come to manage such politically fraught cases as Midyear and Crossfire. Ruefully, I registered the thought that Trump's bullying had broken the system. Instead of fulfilling the roles envisioned by our Founding Fathers, the Republican Congress had become an enforcement arm of Trump's warped cult of personality. That day the results were on display for the American public to see.

When I finally left the hearing that night, I felt relieved. I was glad it was done, and I was glad I had done it. Catching up on the reporting about the testimony on the way back to the FBI, I was gratified to see the number of articles noting my passionate defense of the Bureau. One of the things I noted from the many supportive texts that I received that day, which was also highlighted in the coverage, was the curious and deeply disappointing lack of similar — and more powerful — voices taking up the FBI standard.

My committee testimony had been intended to convince the members of Congress that they needed to keep their eyes on the ball — that is, the threat from Russia, not my texts. But something happened the next day that proved my point better than I ever could have done myself.

The day after my testimony, on Friday, July 13, Rosenstein announced the indictment of 12 Russians accused of interfering in the 2016 elections. I figured that would make for a prickly meeting for Trump, who was headed to Helsinki to meet with Putin. But again I had misjudged the president.

The results of the Helsinki meeting between Trump and Putin a few days later were disastrous, at least for the United States. In the news conference that followed his two-hour private meeting with Putin on July 16 (the subject of which, as of this writing, has not been made public), Trump declined to take Russia to task for meddling in the 2016 elections and repeatedly shifted the blame to his own intelligence agencies and their investigations of that very same interference. Senator John McCain called it "one of the most disgraceful performances by an American president in history," and I couldn't agree more.

As I watched the news conference, I was appalled that Trump had opted yet again to place Putin's empty, self-interested reassurances over the unequivocal conclusions of his own intelligence community. But that professional disgust changed to personal shock when Trump concluded the news conference. In his final remarks, the president shifted from attacking the Russia investigation overall to attacking me — by name.

Towering over the diminutive Putin yet somehow managing to appear subservient, Trump ended the news conference with these words: "And if anybody watched Peter Strzok testify over the last couple of days, and I was in Brussels watching it, it was a disgrace to the FBI. It was a disgrace to our country. And you would say, 'That was a total witch hunt.' Thank you very much, everybody."

I had no idea what Putin must have felt. Part of me wondered darkly if he had a brief moment of panic regarding Trump, as in, *Hey buddy, keep it cool, everyone's going to figure out our secret.* At least Putin had the sense to trust his intelligence agencies, having been a member of one of them. Trump seemed content to eviscerate his — on live television, standing next to the leader of our primary adversary and someone who, as many of us in the U.S. intelligence world now knew, had amassed a staggering amount of coercive leverage over the American president.

THICK LETTER, THIN LETTER

While I sat at home following coverage of Trump in Helsinki on July 16, back at the office the appeal of my proposed termination was working its way up the chain to the head of the FBI's Office of Professional Responsibility (OPR). The assistant director had held that position for a long time, and many agents were afraid of her. To her credit, that was because she was tough, not because she was capricious or political in her decisions. I'd soon have a chance to discover that for myself.

With Aitan at my side, I sat down with the head of OPR on July 24 and made my case to her directly. We argued that I shouldn't be fired, as had been proposed, because the evidence didn't support it. We pointed out facts that the IG had ignored and omitted from his report that not only rebutted his conclusions but supported a different outcome. Furthermore, such a punishment would be inconsistent with past practice, creating an unfair outcome. I put my trust in her to do her job with the consistent integrity that agents grudgingly admitted about her.

She came back with a final decision that overturned the recommendation to fire me — the original proposal that had come down so suspiciously quickly, almost immediately after the publication of the IG report — and instead offered a lesser punishment in a "last-chance agreement." It's called that because it is essentially a contract between the Bureau and the employee in which the employee gives up his or her appeal rights in exchange for the end of the disciplinary process.

I was torn. On the one hand, I thought the proposed solution was far too severe; on the other hand, part of me really wanted just to get back to work — to put this administrative leave behind me, go back out there and investigate the Russians or Chinese or any other threat for which I had spent my life honing my expertise, and retire quietly when I was eligible in a year and a half.

I thought about the offer for a couple of days, and talked it over with family and friends. Then I decided to accept it. I signed the agreement on July 26 and sent it in, ready to get back to work. But as it turned out, it didn't matter, because the FBI leadership reneged on the agreement.

On Thursday, August 9, I was at home when an email from my boss appeared in my in-box. "Are you going to be around tomorrow afternoon or evening? We have a couple of letters to deliver and need to come and see you."

Figuring it was the finalized last-chance agreement, I said I could meet whoever was delivering the letters at my home. I was eager to get all the drama behind me and get back to work.

Since the Helsinki meeting on July 16, Trump had continued stampeding over the special counsel without restraint. If anything, he appeared to be getting bolder. At the beginning of August, he had tweeted that the recused AG — Sessions — should end Mueller's investigation. A week later, Michael Cohen sat down with the special counsel's team and promptly lied about the Moscow project, hoping that the lies would convince Mueller that the project had ended before the Iowa caucus and the first presidential primaries. At that time Cohen was still trying to protect Trump, but the wall was beginning to crumble.

The following morning, August 10, I talked to a colleague in the Human Resources Division about getting all the paperwork delivered. He lived out past my house in the far western D.C. suburbs, so I made plans for him to come to the house with the material in the afternoon in case he wanted to get a jump on the Friday rush hour. When I mentioned the delivery to my wife, she had a sense — a premonition — that perhaps she ought to come home from work. While we waited, we used our nervous energy to neaten up downstairs a bit, rounding up random dog toys and tossing them in a basket. In an effort to be hospitable, we also grabbed some soda from the garage fridge and brought it inside.

Around 4:30 p.m. my colleague showed up. He wasn't alone; he had one of our section chiefs with him, the head of the FBI's recruiting section. I had grown to like the section chief because of his intellect, sense of humor, and enthusiasm for his work. But his presence struck me as odd, adding to a gnawing sense of worry in my mind.

I welcomed them and took them into the living room. We sat down. My colleague had a pained expression on his face as he produced a thick business-sized envelope. *Here's all the last-chance agreement paperwork from OPR,* he said, passing the bundle over. *But here's the letter you want to read,* he added apologetically, handing over a slim envelope, as thin as every rejection letter you ever got from a college or job.

I knew immediately what was in the letter. Blessedly, any emotional response I might otherwise have had was overridden; I felt only the cold detachment of an investigator as I intuited what was inside. I opened the envelope, pulled out the short letter, and read it silently in front of my two colleagues.

The letter was from Bowdich. While OPR had approved the last-chance agreement, he wrote, he was overturning the decision. I was terminated, effective immediately. My FBI career was over. After a quarter of a century in pursuit of the nation's enemies, I had been deemed an enemy myself.

I sat stock-still, silently looking down at the letter and then straight into the uncomfortable eyes of my now former colleagues. A hot gust of anger rushed through my mind and words to go with it crept onto my tongue: *Nice of the stand-up chain of command to have the balls to deliver the bad news themselves.* But I was stubbornly determined not to betray my fury, and I said nothing.

After the slightest pause, I folded up the letter and returned it to the envelope. I stood up at the same time as my former colleagues. *Guys, thanks for coming. I know this wasn't an easy trip,* I said. I could tell it wasn't. They both looked miserable. As we moved toward the door, the section chief broke from his afternoon's assigned role. *I just want you to know, we all watched your testimony, and I couldn't be more proud of how you did,* he said. Then he gave me a bear hug and they walked out the door.

BOXES

June 15, 2018, was my last day as an agent working inside FBI headquarters. I stood in my windowless 10th-floor office, the fluorescent lighting casting a cool blue glow over the room. My boss and another colleague leaned against my desk, all of us awkwardly cracking jokes as I filled up copier-paper boxes with 22 years' worth of memories: photographs, awards, and souvenirs of investigations, gag trinkets, and reminders of both my real family and my FBI family.

Beneath the veneer of joviality, bitterness filled me. The two men were there out of kindness and decency, but I also knew they had to be there to witness my packing, to ensure I didn't try to remove anything I wasn't supposed to. I worked my way down each shelf, picking up mementoes of cases, curiosities from overseas, gifts from colleagues. I carefully placed a small rosewood-colored buddha from Taiwan in a box, and a red-and-white keffiyeh with a tasseled black agal from a trip to the Middle East. Somewhere I stashed my copy of the Putin calendar, like the ones I had handed out to the Crossfire team as gag gifts weeks after the election. And, of course, a Russian matryoshka doll with an espionage theme: a doll portraying a spy with a monocle listening in on a conversation nested inside a doll of an old Russian woman commanding, in Cyrillic, "Keep your mouth shut!," a reproduction of a 1941 Soviet propaganda poster by Nina Vatolina.

When I left the Hoover Building with my belongings, I felt a tremendous sense of sadness. I pulled the minivan into our driveway that night unsure of what I wanted to do with the carload of a career. I decided that I wasn't ready to decide. I went inside to join my wife for a glass of wine and dinner without unpacking the van.

The next day I unloaded everything into our living room, stacking the boxes in a corner and leaning a few framed pictures from the office against them. After a few days of walking past this forlorn monument, I opened the door to our basement, flipped on the light, and walked down the wooden stairs into the still, cool air. I wanted to do this work alone.

I went up and down the stairs, taking the items down one by one and arranging them into a neat stack in a dark subterranean corner. When I had finished, I sized up the dimensions of the pile, then walked over to the area of our basement where I stored my tools. I had folded up a clean plastic tarp to use for a future painting project, but now I had another use for it. Working around the pile, I tucked the edges of the tarp into the seam between the bottom boxes and the floor.

It seemed inconceivable to me that my life's work could all be here, wrapped up under a dustcover, like sheet-draped furniture in an old house. Relics, carefully put into storage in the hope that they might be dusted off at some indeterminate point in the future, when the world righted itself.

Standing there in the half-dark, I thought back to when I was green and just starting out in Boston. A senior agent named Charlie had taken me under his wing. He was an older agent, probably older then than I was now, and hard to read. Back then, I believed that we would always get the bad guys and that justice inevitably rolled down upon the deserving. Charlie cautioned me otherwise. On a gray day, as we drove past row after row of beautiful brownstone homes on Commonwealth Avenue, he suddenly spoke from the passenger seat.

Understand that we don't have a truly equal system of justice, he said. *Some crime is too complex, and with enough money and political clout, the bad guys can bury us, or just wait us out. Worse,* he continued, *with enough power, they can find ways to make the injustice legal. Don't forget that.*

Charlie wasn't despairing about our system of justice when he made those comments. He was tempering my idealism, hoping that I might someday be mature enough to understand the paradox of our responsibility as agents of the Federal Bureau of Investigation: working for justice also entailed perfecting it — just as he and countless agents before him had done.

I stepped back from the pile in our basement, assessing my work, looking for gaps in the tarp. I didn't see any, but I didn't look too hard. I knew the boxes wouldn't remain in the basement forever. My service to the government might have been interrupted, but I couldn't resign myself to this ending. There was still work to be done.

Epilogue

JUST AFTER 7 a.m. on July 25, 2019, President Trump picked up his phone to tweet, yet again, his verdict on the special counsel investigation. This time, at least, the case had reached its natural end.

The day before, Mueller had testified to Congress about his findings. The conclusions of the investigation that I had opened in 2016 were contained in what had come to be known simply as the Mueller Report—an exhaustive, meticulous, and compelling account of the Russian government's multifaceted attack on the 2016 elections, Russia's desire to assist Trump, and his campaign's willingness to accept such help. The report presented a picture of an aspiring president aided by Russia, actively engaged in hiding the truth about those connections, and his efforts to obstruct investigations into that behavior, all the while beholden to others to maintain that deceit. A compromised man.

Noting that he believed existing DOJ policy prohibited the Special Counsel's Office from charging Trump, Mueller had laid out a path forward as plainly as his respect for the institutions of justice would allow. He might as well have said, *Congress, here's a roadmap. It's up to you.*

But none of that mattered to Trump. He waited only to see that no criminal charges would result from the investigation before claiming "complete exoneration." Elated, he tweeted to the world the following morning, "Yesterday changed everything, it really did clear the President. He wins . . . Today, you say impeachment, you have a party of one."

About two hours later, Trump picked up his phone again. This time, though, it was for a call between the White House and Ukrainian president Volodymyr Zelensky. The call lasted about 30 minutes. During the two presidents' conversation, Trump mocked Mueller for his testimony to Congress the previous day, saying "that whole nonsense ended with a very poor performance by a man named Robert Mueller, an incompetent performance."

That was not all. In that same call, the day after Mueller had set forth the results of his work, Trump managed to do what three years of investigation by the FBI and the special counsel had not: generate an impeachment investigation in the House of Representatives.

During their half-hour phone call, Zelensky asked Trump for Javelin antitank missiles to shore up Ukrainian defenses against the Russian incursion in Crimea. Trump immediately responded that "I would like you to do us a favor though." The favor the president wanted was for Zelensky to work with Trump's new attorney general, William Barr, and Trump's personal attorney Rudy Giuliani to dig up information about a long-debunked theory that Ukraine, not Russia, was behind the election-hacking activity in 2016. He also wanted Zelensky to investigate one of his Democratic political rivals, former vice president Joe Biden, and his son.

"Our country has been through a lot and Ukraine knows a lot about it," Trump said to Zelensky. Shortly after, he added, in reference to the 2016 election interference and the investigation that followed, "They say a lot of it started with Ukraine." He was speaking, of course, about the case my colleagues and I had begun pursuing before the election — a case whose voluminous, airtight conclusions had been discussed on Capitol Hill at great length the previous day.

The Ukraine meddling story was a hoax, spun up in the conspiratorial corners of right-wing and pro-Russia websites and almost certainly propagated, if not created, by Putin. It was perfect Russian propaganda: deflect attention and blame for Russian behavior onto Ukraine, all the while throwing discord into the domestic U.S. political debate. But Trump, obsessed with undercutting the intelligence community consensus about the meddling Russia had done on his behalf, didn't care. After a campaign in which Russian disinformation flooded the Internet and social media, he was, effectively, trying to generate disinformation of his own: an active measure directed against Americans by their own president and aimed at keeping Trump in the Oval Office.

Unchastened by Mueller and worried about losing to Obama's vice president, Trump had waited less than 24 hours to seek foreign help to win another presidential election.

I recite these familiar details to make a simple point: the Russia and Ukraine investigations are part of the same story. Since the very start of the impeachment inquiry, Trump's supporters have derisively argued that because the special counsel investigation failed to directly implicate President Trump in the Russians' election meddling, his critics — seeking a new line of attack — instead seized on his dealings with Ukraine. But Ukraine and Russia are not separate issues; they are points on the same long arc bending toward corruption.

Ukraine is woven throughout Trump's preelection narrative in both 2016 and 2020. The only real difference is that in 2019, the compromised individual was not a long-shot presidential candidate but rather the chief executive of the United States, a man who could wield the full power of his office for personal gain — and whose malfeasance also implicated, and compromised, a much larger organization than a presidential campaign. The fundamental pattern, however, endured: Paul Manafort's support of pro-Russian Ukrainian president Viktor Yanukovych; the 2016 Republican Party platform change removing policy support for lethal defensive military aid to Ukraine; Giuliani's championing of a slowly simmering false narrative that Ukraine, not Russia, was behind foreign interference in the 2016 elections. All of these actions by people or organizations closely associated with Trump hurt the U.S. and our allies while serving Russian interests by hiding or clouding Russian involvement. Studying this grimy mosaic reveals campaign managers, campaign advisers, donors, attorneys, and even Trump himself.

The political din around the Mueller Report drowned out the fact that it confirmed and deepened America's understanding about ongoing Russian active measures, Russia's attack upon our democracy in 2016, and the enthusiastic embrace of those efforts by people around Trump, including his son. Punctuating the campaign's clear efforts to capitalize on the Russian measures, Trump's informal adviser Roger Stone, a self-described dirty trickster, was convicted on November 16, 2019, on seven felony charges for trying to sabotage a congressional investigation into Russian election meddling. Stone attempted to serve as a back channel between WikiLeaks and the campaign, a role in which he worked to coordinate and capitalize on the release of the stolen Democratic email, like a drug dealer attempting to pump toxins straight into the veins of the body politic. Like Manafort, like Gates, like Papadopoulos, like Flynn, he got caught, and — as of this writing — faces imprisonment.

Two days after Mueller first submitted his report to DOJ, on March 24, 2019, Barr sent to Congress — and released to the public — a summary of Mueller's report that seriously distorted the special counsel's findings and mystified those of us who had worked on the Russia investigation. Barr claimed, among other things, that "the Special Counsel's investigation did not find that the Trump campaign or anyone associated with it conspired or coordinated with Russia in its efforts to influence the 2016 U.S. presidential election." I had spent a year uncovering the breathtaking efforts of the Trump camp to connect with and benefit from the Russians' attacks on our democracy, and Barr's statement seemed like nothing less than a malicious fabrication intended to protect the president at the expense of truth and the republic.

Disappointingly, Attorney General Barr's characterizations of Mueller's investigation were at odds with the damning nature of the report. They seemed to be cynical efforts to blunt the report's impact by seizing the narrative and replacing it with a *1984* version. As George Orwell wrote, "If all others accepted the lie which the Party imposed — if all records told the same tale — then the lie passed into history and became truth." Barr was either a student of Orwell's or was unconsciously following the authoritarian playbook the British writer described in his novel, because his initial response to the report had "Big Brother" written all over it. Barr's characterization of the Mueller Report, indeed, was so disingenuous that Mueller subsequently wrote a letter to Barr expressing his concern that Barr's summary "did not fully capture the context, nature, and substance" of Mueller's work.

Coming from Mueller, a man whose every fiber is imbued with respect for the structure and constraints of the institutions of justice, this formal written objection was stunning — a screaming klaxon warning Americans that truth was being subverted. In the congressional testimony that followed, Mueller exposed Barr's anodyne assertions about the special counsel's conclusions for the mischaracterizations they were. But still, it wasn't enough to undo the damage that the attorney general had done.

The problem was, it wasn't easy to discern Mueller's actual conclusions from his testimony. The format of congressional testimony, the five-minute blocks of helter-skelter questions crafted to obfuscate and confuse, muddled Mueller's explanation. The structure was one of the worst formats for the reserved Mueller to make his case about the significant yet complex conclusions of the report. The president's champions capitalized on the complexity of the case and the challenge of distilling it into sound bites to throw its findings into doubt. They might as well have taken a page from a Russian playbook of active measures. Overload the debate with competing theories. Create false equivalences. Raise doubts and fan suspicion. Divert attention and sow division. But at all costs, avoid the truth.

Within this morass, Russian intelligence services have had a field day. It's the perfect environment for *state-level* dirty tricksters to deceive and coerce. And just as Trump's unacknowledged Trump Tower Moscow negotiations gave the Russians leverage over him, his secret and apparent threat to withhold aid to Ukraine in exchange for personal political benefit gave that same leverage not only to Ukraine but also to any other countries whose intelligence services were likely listening in on the calls and discussions about the White House's shadow campaign. Trump's efforts in Ukraine compromised him yet again, and made him vulnerable to coercion by yet one more foreign power, if not more.

The only antidote to secret misconduct is transparency, and there are signs that it is growing. Several months after Trump's July call with Zelen-sky, a whistleblower came forward, detailing the administration's extortion of Ukraine to obtain personal political benefit in exchange for the release of military aid that would enable Ukraine to defend itself against Russian in-cursions into its sovereign territory. It is hard to overstate the importance of this brave individual, whose action ultimately triggered the impeachment investigation into President Trump.

The whistleblower's complaint brought forth a series of professional dip-lomats and military officers who corroborated the account of Trump's and others' behavior, placing themselves at risk in order to honor their duty to their oath of office. But it also highlighted the chilling effect of Trump's attacks on the apolitical government bureaucracy: dozens of people were aware of the administration's attempts to condition military aid for Trump's personal purposes. If not for the whistleblower complaint, it's possible that none of these officials would have come forward publicly.

I can't judge them too harshly. Having experienced the searing focus of a Trump-led effort to vilify and turn me into a false caricature of bias and incompetence, I watched as the familiar spotlight turned its harsh glare upon the men and women drawn into the impeachment inquiry. I watched as partisans across the country, and at all levels of government, sought to denigrate these public servants, distorting or ignoring facts to impugn their characters and motivations.

That modus operandi is also Trump's. Raise the personal and profes-sional costs of dissent. Rob federal agencies of the ability to defend them-selves and their employees. Wield fear openly, brazenly treating such be-havior as normal and daring any critic or branch of government to call out its impropriety.

When I look at the impeachment witnesses, I see that most of them are men and women trying their best to uphold their oaths to the Constitution, doing everything in their power to honor the trust and power placed in their hands by the American people. I see patriotism, I see humility, and I see duty to America. I see what I saw in all of us at the FBI as we responded to the Russian attacks on the 2016 election.

But even well-intentioned government workers speaking truth to power have little chance when that power is an unbridled American presidency. And just as Trump's statements cannot fail to have an impact on our current public servants, his words and deeds also hamper our preparedness for future election interference.

Every FBI agent and analyst I know is steadfastly dedicated to protecting and defending the Constitution and the United States, yet Trump has repeatedly brought his weight to bear against efforts of the FBI and the U.S. intelligence community to combat malign foreign influence on our upcoming elections. Since he first called for Russia to weaponize Hillary Clinton's stolen email in the run-up to the 2016 election, he has continued — in public and in private — to solicit foreign interference in the most sacred mechanism of our democracy. With Congress demonstrating daily that it has no appetite to address the threat and the president lying brazenly about those of us who handled the last investigation, ask yourself this question: if you were an FBI agent today, would you be looking to stick out your neck to show that Russia aims to repeat in 2020 what it did in 2016 — once again with Trump's explicit encouragement? I know that despite the fear, agents and analysts are nonetheless pursuing the truth, and I have nothing but the highest respect for their professionalism and courage. But I can only imagine how different their work might be in a different environment. This is far from the only "what if?" that I permit myself these days. Such opportunities for reflection present themselves relentlessly.

On December 11, 2019, the DOJ inspector general finally released his report on the beginnings of Crossfire Hurricane. Horowitz's conclusion was ultimately the same as my reaction when his investigators had first called me at the Special Counsel's Office 28 months earlier: of course my personal, private political beliefs hadn't affected what I had done. In the course of investigating both Crossfire and Midyear, the IG had conducted detailed, years-long examinations of every email, every text, every note, and all of my conduct. The IG staff had reviewed thousands upon thousands of investigative documents. They had interviewed hundreds of people, specifically asking them if they had ever detected any bias or improper political considerations in me or my work. Add to the IG's work the investigations conducted by the House and Senate Judiciary and Intelligence Committees, as well as unending digging by the media. They found the truth: my personal beliefs never influenced my work.

I have a bitter taste in my mouth when I consider what the world might look like if DOJ had not illegally disclosed my texts, mid-investigation, to the media. I wonder how things might be different if the inspector general —while acknowledging in congressional testimony that his Crossfire investigation had not even gotten off the ground—had not voiced concern about a speculative possibility, going beyond his own findings to provide grist to the partisan mill by claiming that he "couldn't rule out" that I might have acted with bias in opening Crossfire before he had yet to investigate that supposition. He claimed he couldn't prove a negative, but once he took the time to properly review the evidence, he did, concluding that there was no "documentary or testimonial evidence that political bias or improper motivation influenced the decisions" to open the cases.

I try to understand why, during that testimony, the inspector general engaged in the very behavior he was simultaneously castigating Comey about, for "deviating from well-established Department policies" by engaging in what felt like character assassination in the middle of an investigation.

I try to imagine what the national political dialogue would have been if the normal, proper process had been followed—if the first anyone heard of me was a report in December 2019 stating that the IG had identified personal text messages of a political nature but had concluded that my personal beliefs had not had any impact on my professional acts. I wonder how Mueller's efforts would be viewed differently, and the other ways in which the various investigations might have unfolded. I wonder whether the proposition that all U.S. citizens have a duty and obligation to be engaged in our civic discourse—that government employees can and should have political opinions, that they usually do, and that that doesn't affect the objectivity with which they approach their jobs—would seem quite as objectionable to as many Americans as it does today. All that wondering adds tremendous sadness to my bitterness.

So why write this book? Why solicit more assaults from critics — many in the highest parts of our government — who have deliberately attempted to distract attention from the real issue of a compromised president's corrupt complicity in Russian interference? Especially when, aside from required testimony, I have stayed silent for years, first out of duty to the FBI's rules and then to protect and respect an ongoing investigation?

Because the Russians haven't gone away.

Mueller may have shuttered his office. DOJ may have declined various prosecutions (although, thankfully, several investigations continue). Men and women vigilant to national security threats may have been chased out of public service. But the Russians are there. And they're coming at our 2020 elections with a vengeance.

On the morning after Election Day in 2016, several of us were concerned that the Russians had pulled back from plans to deploy disruptive actions in the event of a Clinton victory. We suspected that Moscow had shelved those operations because Trump had prevailed but were keeping them ready for use in 2020, all the while developing new avenues of attack.

This is still my fear. I expect that future elections will see the Russians engaging in all the sorts of active measures we saw in 2016, and that they'll be bringing new tactics to the fight as well: Altering voter rolls. Tampering with voting results, in however limited a way, and amplifying the news of that meddling on social media. Hacking into and crashing voting infrastructure. Spreading false stories of disenfranchisement and voting fraud. Releasing *kompromat* that Russia has spent years collecting and that has the potential to be greatly disruptive.

We'll find out soon enough.

I'm speaking out because we all need to speak out, to do the right thing, and to join the fight publicly. Now that I'm no longer in the FBI and Mueller's work is done, I'm free to do so in public as well as in private. Free expression of this kind is what America is all about.

So is the freedom to choose our own leaders. Indeed, our national identity is defined by the free and fair election of our representatives, up to and including the person who occupies the highest office in the land. Foreign nations have no place in that process, but Russia has proven determined to insert itself into it.

Our investigations revealed Donald Trump's willingness to further the malign interests of one of our most formidable adversaries, apparently for his own personal gain. They also showed his willingness to accept political assistance from an opponent like Russia — and, it follows, his willingness to subvert everything that America stands for.

That's not patriotic; it's the opposite.

The Founders gave us a remedy for a leader such as Trump, who elevates his personal interests above his public duties: impeachment and removal from office. Failing that, we must rely on the ballot to judge him. But if things continue as they have under our compromised 45th president, I fear the judgment will be on us for our failure to uphold America's core values. And Russia will reap the rewards.

Acknowledgments

During the early stages of this book, I expected the process to be cathartic, but I did not anticipate how difficult the work proved to be. As someone prescient noted to me — *who told you catharsis is supposed to be easy?* — writing proved to be an act of discovery as much as remembrance. The journey renewed my appreciation of the blessing of family and friends. I would not be here without the love, support, and good counsel of my wife and children. Your grace, strength, resilience, and sense of humor are remarkable, and I treasure you more than can be expressed.

To Mom and Dad, thank you for simultaneously holding admiration and respect for the diversity of the world while teaching me the exceptionalism of what America strives to be, for living an example of public service and adventure, and encouraging me on my own path toward that ideal. To our extended family and friends who stood with us during even the most outrageous events, thank you — your support, time, understanding, and hospitality were and remain a godsend.

This book could be filled with scores more remarkable stories of the men and women of the FBI. Thank you for your friendship and support. Your patriotism and duty inspire me, and it is my honor to know and have worked with you. To the prosecutors, intelligence professionals, and foreign colleagues with whom I have had the privilege to work, thank you. I will always admire your expertise, hard work, and shared passion for our common mission.

Other true professionals made this book possible. To my literary agents Gail Ross and Howard Yoon, and to my various named and unnamed editors, including Alex Littlefield and Bruce Nichols and the entire team at Houghton Mifflin Harcourt, thank you for your comments, suggestions, and everything you did to improve this book. You made it significantly better — the flaws in it are mine alone. I will forever be indebted to the legal acumen of Aitan Goelman, his colleagues Kate Duval and the inimitable Roger Zuckerman, as well as Jim Sharp and Steve Grafman. The dogged professionals at the Knight First Amendment Institute were critical to getting my manuscript through a very long-running prepublication review process; you have my gratitude. To other unnamed but extraordinary attorneys who provided their wisdom, advice, and expertise — thank you.

To everyone who took the time to write, call, email, or tweet, your support means the world to me — it gave comfort in difficult times and hope in moments of doubt. Each kind word, donation to my legal fund, anonymous picking up of a lunch tab, and other countless displays of kindness mattered. You embody the fundamental decency of the American people. Your actions humble me and serve as reassurance that we will always rise to the challenge before us. Thank you.